weird earth

weird earth

DEBUNKING STRANGE IDEAS
ABOUT OUR PLANET

DONALD R. PROTHERO

RED ⚡ LIGHTNING BOOKS

This book is a publication of

Red Lightning Books
1320 East 10th Street
Bloomington, Indiana 47405 USA

redlightningbooks.com

Manufactured in the United States of America

First printing 2020

ISBN 978-1-68435-061-2 (hdbk.)
ISBN 978-1-68435-136-7 (web PDF)

*This book is dedicated to the great geologists
who trained me and inspired me:*

Harry Cook

Mike Woodburne

Mike Murphy

Lewis Cohen

Peter Sadler

Paul Robinson

Wally Broecker

Neil Opdyke

Dennis Kent

Bill Ryan

Walter Pitman

Rich Schweickert

Larry DeMott

Dewey Moore

Bob Dott

CONTENTS

FOREWORD:
ROMANCING THE STONE

The book you hold in your hands by my friend and colleague Donald Prothero is one of the most captivating you will ever read. Once you start in, you won't be able to put it down as you will be constantly amazed by what strange ideas people have about our planet. I've been studying weird beliefs for over a quarter century, and in reading this book I was still stunned by what some members of my species think about earth, including that it is at the center of the universe, that it is only six thousand years old, that all those dinosaur fossils are faked, that it is a giant magnet, that it is flat, that it is hollow, that it is constantly expanding, that we never visited its moon, that there are mysterious ley lines around it directing the planet's energies, and that there was once an ancient advanced civilization on it called Atlantis.

On this last claim, on May 16, 2017, I spent nearly four hours on Joe Rogan's wildly popular podcast debating an alternative archaeologist named Graham Hancock, who believes that long before ancient Mesopotamia, Babylonia, and Egypt there existed an even more glorious civilization that was so thoroughly wiped out by a comet strike around twelve thousand years ago that nearly all evidence of its existence vanished, leaving only the faintest of traces that he thinks include a cryptic warning that such a celestial catastrophe could happen to us.

Hancock has put forth variations on this general theme in numerous well-written and best-selling books, including *Fingerprints of the Gods: The Evidence of Earth's Lost Civilization* (1995), *The Message of the Sphinx: A Quest for the Hidden Legacy of Mankind* (1997), *Underworld: The Mysterious Origins of Civilization* (2002), *Magicians of the Gods* (2015), and most recently *America Before: The Key to Earth's Lost Civilization* (2019). I listened to the audio editions of *Magicians of the Gods* and *America Before*, both read by the author, whose British accent and breathless revelatory storytelling style is, I confess, compelling. But is it true? I'm skeptical. As I explained in my June 2017 column in *Scientific American*: "First, no matter how devastating an extraterrestrial impact might be, are we to believe that after centuries of flourishing every last tool, potshard, article of clothing, and, presumably from an advanced civilization, writing, metallurgy, and other technologies—not to mention their trash—was erased? Inconceivable."[1]

Second, Hancock's impact hypothesis comes from scientists who first proposed it in 2007 as an explanation for the North American megafaunal extinction around that time and has been the subject of vigorous scientific debate. It has not fared well. In addition to the lack of any impact craters dated to around that time anywhere in the world, the radiocarbon dates of the layer of carbon, soot, charcoal, nanodiamonds, microspherules, and iridium, asserted to have been the result of this catastrophic event, vary widely before and after the megafaunal extinction, anywhere from ten thousand to fourteen thousand years ago. Furthermore, although thirty-seven mammal species went extinct in North America (while most other species survived and flourished), at the same time fifty-two mammal genera went extinct in South America, presumably not caused by the impact. These extinctions, in fact, were timed with human arrival, thereby supporting the more widely accepted overhunting hypothesis.

Third, Hancock grounds his case primarily in the *argument from ignorance* (since scientists cannot explain X, then Y is a legitimate theory) or the *argument from personal incredulity* (because *I* cannot explain X, then my Y theory is valid). These are "God of the Gaps"–type approaches that creationists employ, only in Hancock's case the gods are the "Magicians" who brought us civilization. The problem here is twofold: (1) scientists *do* have good explanations for Hancock's Xs (e.g., the pyramids, the Sphinx), even if they are not in total agreement, and (2) ultimately one's theory must rest on *positive* evidence in favor of it, not just *negative* evidence against accepted theories.

Hancock's biggest X is Göbekli Tepe in Turkey, with its megalithic T-shaped seven- to ten-ton stone pillars cut and hauled from limestone quarries and dated to around eleven thousand years ago when humans lived as hunter-gatherers without, presumably, the know-how, skills, and labor to produce them. Ergo, Hancock concludes, "At the very least it would mean that some as yet unknown and unidentified people somewhere in the world

---

1. Michael Shermer, "Romance of the Vanished Past," *Scientific American* 317, no. 6 (2017): 75.

had already mastered all the arts and attributes of a high civilization more than twelve thousand years ago in the depths of the last Ice Age and sent out emissaries around the world to spread the benefits of their knowledge."[2] This sounds romantic, but it is the bigotry of low expectations. Who's to say what hunter-gatherers are or are not capable of doing? Plus, Göbekli Tepe was a ceremonial religious site, not a city, as there is no evidence that anyone lived there. Furthermore, there are no domesticated animal bones, no metal tools, no inscriptions or writing, and not even pottery—all products that much later "high civilizations" produced.

Fourth, Hancock has spent decades in his vision quest to find the sages who brought us civilization. Yet, decades of searching have failed to produce enough evidence to convince archaeologists that the standard timeline of human history needs major revision. Hancock's plaint is that mainstream science is stuck in a uniformitarianism model of slow, gradual change and so cannot accept a catastrophic explanation. Not true. From the origin of the universe (Big Bang), to the origin of the moon (big collision), to the origin of lunar craters (meteor strikes), to the demise of the dinosaurs (asteroid impact), to the numerous sudden downfalls of civilizations documented by Jared Diamond in his book *Collapse*, catastrophism is alive and well in mainstream science.

The real magicians are the scientists who have worked this all out.

On this final point about scientists as magicians: as someone with zero training in geology, when I read about the work that professional geologists like Donald Prothero have conducted to determine the age, nature, and processes of the earth, I feel exactly the same way I do as when I see the magicians Penn and Teller catch bullets in their teeth or when the magician David Copperfield makes the Statue of Liberty disappear. When you don't know how the trick is done—or in this metaphor, how the science is conducted, understanding, say, how geologists determined that the earth is 4.6 billion years old—it feels like magic to me. But once geologists like the author of this book reveal how the secret of science is done, you understand that it's not real magic, as in paranormal or supernatural forces at work. It's scientific magic.

Romancing the stone we call earth through all these alternative theories may appeal to our fantasies and imaginations, but ultimately we want

---

2. G. Hancock, *Magicians of the Gods* (New York: Griffin, 2015), 32.

to know what is true. So as you read *Weird Earth*, I hope the scales will fall from your eyes as they did mine when I read it, understanding fully why all those crazy ideas about our planet that people have concocted over the millennia are wrong and why science really is the best tool we have for understanding nature.

MICHAEL SHERMER is the publisher of *Skeptic* magazine, the host of the Science Salon podcast, and a Presidential Fellow at Chapman University. For eighteen years he was a monthly columnist at *Scientific American,* and he is the author of a number of *New York Times* best-selling books, including *Why People Believe Weird Things, The Believing Brain, Why Darwin Matters, Heavens on Earth,* and *Giving the Devil His Due.*

PREFACE

Having written a book debunking UFOs and aliens (*UFOs, Chemtrails, and Aliens: What Science Says*, 2017, with Timothy Callahan) and also a book about cryptozoology (Bigfoot, Nessie, the Yeti, etc.) titled *Abominable Science* (2013, with Daniel Loxton), it occurred to me that there needed to be another book about the huge number of weird, paranormal, and supernatural ideas people hold about the earth. They range from crank ideas about geology (hollow earth, flat earth, geocentrism, moon-landing hoaxes, expanding earth, myths about the earth's magnetic field), to mystical and paranormal explanations of earth features (aliens at Mount Shasta, ley lines, Atlantis, Lemuria), to mystical and nonsensical ideas about natural objects and processes (crystal healing, dowsing). As a professional geologist with forty years of experience teaching college geology, I have a broad background in many topics in the earth sciences, so I can discuss the reasons why these weird ideas are not real.

Since 2011, I have been writing almost-weekly blog posts on www.skepticblog.org about a wide variety of topics in science and skepticism, and I have often found myself addressing popular nonsense about the earth. I thought it would be a good idea to collect all these weird ideas into a book so that their common background and threads could be examined.

Thus, this book covers most of the weird ideas that have been promoted concerning our planet, especially the major ones like flat-earth beliefs, geocentrism, expanding earth, hollow earth, and the myths propagated by the Young-Earth Creationists. In addition, it covers some of the minor ones that most people haven't heard of but that are popular in the paranormal and New Age communities of believers, such as the aliens in Mount Shasta, ley lines, and crystal healing. In addition to these individual topics, the first chapter provides a background to how science works and how scientists think so that we can better understand why science rejects these weird ideas.

And the last chapter looks at the psychology of why people believe these weird things and what it implies for our future.

Of course, this book could be twice as long if I included nearly every topic that touches the earth in some way. There are lots of weird ideas about the weather (such as UFO clouds and chemtrails) that I have covered elsewhere, as well as auroras, ball lightning, the Tunguska explosion, and even conspiracies about humans controlling the weather, but these are largely outside the domain of geology. There are myths about aliens producing features on the landscape or creating crop circles, the Planet X/Nibiru idea, but those are largely covered in my book on UFOs and aliens.

Then there is the entire domain of apocalyptic beliefs and legends, and wild ideas about the end of the world, but debunking these is largely the domain of deconstructing religious prophecies and mistranslations of ancient Mayan inscriptions (as in the supposed end of the world back in 2000). These are not really about geology. Whenever these end-of-the-world scenarios mention actual geologic events, it's just a mishmash of all sorts of natural disasters (earthquakes, tsunamis, volcanoes, floods, storms) thrown together as a mechanism to end the world, not a serious misunderstanding of how these processes work.

In many of the chapters, I've tried to go beyond just debunking the claims and have added a section called "How Do We Know?" that outlines the scientific evidence for *why* science rejects the ideas in this book as crackpot and wrong. I feel that this is one of the most important things that a reader can take away from a book like this: not only knowing what is false but also understanding *why* we know it is false. Otherwise, much of what people learn about science is just memorizing facts and dogmatic assertions (a common problem in our K–12 science educations); as a result, students (and readers of this book) don't understand or even hear the huge amount of evidence for *why* science rejects some ideas and accepts others. I hope readers will realize and appreciate that this is the most important thing they can learn from a book like this.

ACKNOWLEDGMENTS

I thank Ashley Runyon and Rachel Erin Rosolina at Indiana University Press for their interest and support of this project and Carol McGillivray for producing this book. I thank Sharon Hill and another anonymous reviewer for providing careful reviews of the chapters. I thank my wonderful sons, Erik, Zachary, and Gabriel, and my amazing wife, Dr. Teresa LeVelle, for their support and forbearance as I disappeared into my office for the entire Christmas vacation of 2018 to write this book.

weird earth

①

Science and Critical Thinking

The Science of the Earth

Like it or not, we all live in an age of science and technology. Science has utterly transformed the lives and fates not just of humans but of all organisms on the planet. Just look at what science has given us. Only 150 years ago, most children died in childbirth or through incurable childhood diseases. Today, thanks to modern medicine, nearly all children in the developed world survive their birth and early years. We take for granted that most of us carry a device in our pocket that is more powerful than a room-sized computer from only fifty years ago; it also performs as a phone, pager, clock, calculator, and video and audio player and has many other functions. Until the invention of the steam locomotive and then even faster transports, no human could travel any faster than a horse could run. Now we all routinely travel at 65 miles per hour on highways, and many people have flown and traveled faster than sound. Our lives are so completely dependent on the miracles of science and technology that we don't even think about them anymore. We are aware of our dependence on them only when we lose them, such as during a power outage or an earthquake or other natural disaster.

Likewise, over the past two hundred years, the scientific method has been applied to the study of the earth, and its progress has led to great discoveries. We now know of millions of extinct animals that lived long before humans ever appeared. We can date rocks with high precision and can estimate the age of the origin of the earth and solar system at 4.56 billion years. We know what shapes the surface of the earth, what is beneath the surface, and how continents move around the earth's surface. Instead of viewing earthquakes as a sign of the wrath of the gods, we understand what causes them and have made enormous strides in understanding and preparing for them, if not predicting them. Modern society runs on coal, oil, natural gas, and uranium, as well as valuable materials like gold, silver, copper, and

1

platinum, and it depends on resources like steel, stone, and concrete. These discoveries and technologies were made possible only by the application of the scientific method to earth sciences by scientists curious to know how the earth worked.

We are completely and utterly dependent on science and technology for our survival, yet we find that even in the most developed countries of the world, a significant number of people reject some aspect of science because it conflicts with deeply held beliefs. They love what science gives them (such as health, technology, and wealth) but reject science when it tells them something they don't want to hear. But we don't get to make that choice. Science is not a restaurant menu that you can pick and choose from. As science educator Bill Nye said, "The natural world is a package deal; you don't get to select the facts you like and which you don't."[1] Or as astrophysicist Neil deGrasse Tyson said, "When different experiments give you the same result, it is no longer subject to your opinion. That's the great thing about science. It's true, whether or not you believe in it. That's why it works."[2]

This is particularly true when science finds out something that goes against what we want to believe—what Al Gore aptly called "inconvenient truths." Scientists don't get to pick and choose what they want to believe when they are doing research. They are obligated by their training as scientists to report their results, no matter how much it might go against what they wish to be true. Science tells us that we are a product of evolution and that we are closely related to the apes, that humans are insignificant on the scale of the cosmos or in the framework of geologic time, and that humans are destroying the planet through pollution and especially climate change. These things are not comfortable or easy to live with and may be a blow to our notions of cosmic importance—but they are true because that's what the evidence shows.

Scientists are not spoilsports or killjoys, and we don't take pleasure in shattering illusions. Despite what some science deniers claim, there's no incentive for scientists to tell you bad news. We don't get more grant dollars for telling you the grim truth about climate change or discovering more evidence of your close relationship to the apes. If a scientist tells you an "inconvenient truth," it is because a scientist *must* do so as a part of honest, objective reporting of what the data show. An amusing online cartoon shows a variety of scientists speaking inconvenient truths and being punished for it—from Archimedes being killed by the Roman soldier as he did his geometry, to Bruno being burned at the stake for saying the earth is not the center of the universe, to Darwinian evolution, to Einsteinian relativity. The final panel says, "Science: if you ain't pissin' people off, you ain't doing it right."[3]

What Is Science?

Science is essential to our daily lives now, but very few people actually understand what it is or how it works. The media feed us a diet of stereotypes, especially the classic "mad scientist" trope, complete with the white lab coat, the sparking apparatuses and bubbling beakers, wild hair, and maniacal laugh. But most scientists don't wear white lab coats. I haven't worn one since I took chemistry lab in college, and the only scientists who need them are those who work with stuff that might splash on their clothes, such as chemists and medical personnel. No, scientists aren't defined by the color of the coat they wear or the gizmos they work with. They are defined by what is in their heads and how they think.

Science is a way of thinking about the world, not how you dress or what toys you play with. Science is thinking critically about phenomena in the natural world and trying to find ways to test hypotheses, or preliminary explanations, about how the world works. As the philosopher George Santayana wrote, "Science is nothing but developed perception, interpreted intent, common sense rounded out and minutely articulated."[4] All science is about testing hypotheses and finding out their validity by further observations and experiments. Scientists generally aren't trying to prove their hypotheses but to disprove them. As British philosopher Sir Karl Popper pointed out many years ago, it's far easier to prove a hypothesis wrong (falsify it) than it is to prove it right (verify it). The famous example is the classic philosophical statement "All swans are white." No number of white swans proves that statement true, but a single nonwhite swan proves it false. Indeed, there are black swans in Australia (fig. 1.1). If your hypothesis has been tested and found false, you must abandon it and move on to another explanation—perhaps one suggested by your previous failure. Popper titled one of his books *Conjectures and Refutations*, a nice summary of the scientific method in a single phrase.

This idea surprises a lot of people, but it is true. Strictly speaking, science is about proving ideas wrong and moving on, not proving them right. Scientists are *not* looking for "final truth" or proving something "absolutely

Figure 1.1. Not all swans are white. This is the Australian black swan. (*Courtesy Wikimedia Commons.*)

true." Scientific explanations must always be open to further scrutiny and testing; they are tentative and must be capable of being rejected. As the famous philosopher Bertrand Russell wrote, "It is not what the man of science believes that distinguishes him, but how and why he believes it. His beliefs are tentative, not dogmatic; they are based on evidence, not on authority or intuition."[5] Whether religious, political, or social, ideas that cannot be tested are not scientific; they are dogma. This immediately distinguishes science from many other areas of human thought. For example, we might say that "Zeus caused the lightning and thunder," but this is a religious belief. It is not a testable scientific idea. Marxism and many other dogmatic worldviews also make broad statements about the world that cannot be tested but are articles of faith among the believers, so nothing would ever prove them false. When dogmatists (religious or otherwise) have their sacrosanct ideas challenged, they will not admit that the idea has been falsified. They stubbornly insist they are right, or they find some dodge to salvage at least some of their false notions.

Thus, science is very different from what most people think it is. When scientists speak to each other, they are not after "truth." They are careful not to use the words *true* or *fact*, and strictly speaking, we don't "prove things true." Instead, scientists are trying to test and falsify, and test again, until an idea is *well corroborated* (not "proven true"). What most people would call a "fact" is an "extremely well-supported explanation." To a scientist, the highest form of a corroborated hypothesis is a *theory*, a group of interrelated and well-corroborated hypotheses and observations that have received widespread acceptance because they explain so much.

Sadly, the public uses these words and concepts very differently. In everyday usage, *theory* means a wild speculative idea, like "theories of why JFK was assassinated." Creationists take advantage of the confusion and exploit this meaning of the word by denigrating evolution as "just a theory." Well, gravity is just a theory too, but the objects around you are not floating around in the air. Thanks to the germ theory of disease, we believe that bacteria and viruses are the major causes of diseases, not some sort of "ill humor" in your blood that your doctor would remove by bleeding you with leeches.

Likewise, in the public debate about scientific topics, science deniers will put down an idea they oppose (like climate change) by saying that it's not "proven true" or "100 percent true." *Nothing* in science is "proven true," and everything has probabilities associated with it. I can't say that I can "prove" you would die if you jumped off a twenty-story building, but I can say that it's likely to happen with a 99 percent probability—and most nonsuicidal people will not take that less than 1 percent chance that they won't die.

As Carl Sagan said, "Skeptical scrutiny is the means, in both science and religion, by which deep thoughts can be winnowed from deep nonsense."[6] Science is basically applied skepticism. We try to be skeptical of all ideas until they have been tested and corroborated again and again, and then we only give our provisional assent. We don't *believe* in an idea; we *accept* it based on evidence. (*Believe* is a religious and cultural word, not a scientific one.) Most humans are cautious of people trying to sell them worthless junk or politicians making impractical promises or swindlers trying to con them into believing something or buying something. We all know that advertising is exaggerated or deceptive or distorted, and in many cases, it is an outright lie. We try to look for good products and avoid junk when we are shopping, and we employ the old Latin maxim *caveat emptor*, "let the buyer beware." Yet many people won't employ the same skepticism to outlandish claims about religious miracles or UFOs or Bigfoot or a wide variety of paranormal ideas that sucker people every day. Most of the ideas in this book fall within the realm of outlandish and even bizarre, but there are plenty of believers. Yet these same people are skeptical elsewhere in their lives and won't fall for a deceptive ad on TV or the internet or a telemarketer trying to sell them something.

Scientists are humans too, and although they try to be hard-boiled skeptics, they cannot avoid falling for the traps in thinking and sometimes embrace ideas that fit what they want to believe rather than what is. As Carl Sagan wrote, "There are many hypotheses in science which are wrong. That's perfectly all right; they're the aperture to finding out what's right. Science is a self-correcting process. To be accepted, new ideas must survive the most rigorous standards of evidence and scrutiny."[7] For this reason, there is an important quality control mechanism built into the fabric of science: *peer review*. This is very different from the internet, which is a giant cesspool of garbage and bad ideas with no fact-checking, and it is very different from partisan media outlets, which have given up reporting anything "fair and balanced" but churn out nonstop propaganda.

Scientists, on the other hand, must submit their ideas to the harsh review and scrutiny of other scientists before they can be published. Usually these reviews are anonymous, and they can be sent to any qualified scientist, including your worst critic. If your idea is rejected, you can give up, or you can try to do a better job of supporting your hypothesis and submit it again. Peer review weeds out the bad ideas in science, and after a harsh round of review before publication, and an even harsher scrutiny in the years after publication, most ideas in science that have survived many years are probably true and have passed quality control.

Peer review is particularly important in evaluating our own ideas, since we are inclined to think our own ideas are right and cannot judge them critically. As the Nobel Prize–winning Caltech physicist Richard Feynman said, "The first principle is that you must not fool yourself and you are the easiest person to fool."[8] Many scientific experiments are run by the double-blind method, in which neither the subjects of the experiments nor the investigators know what is in sample A or sample B. In a double-blind experiment, the samples are coded so that no one knows what is in each sample, and only after the experiment is over do the scientists find out whether the results agree with their expectations or not. As Feynman said, "It doesn't matter how beautiful your theory is, it doesn't matter how smart you are. If it doesn't agree with experiment, it's wrong."[9] Ultimately, bad ideas are weeded out, and good ones survive to become the established framework of scientific theory that all scientists build upon.

The mad scientist stereotype that prevails in nearly all media is completely wrong not just because of the clothing, behavior, and apparatuses that are shown. It's wrong because the "mad scientist" is not testing hypotheses about nature or experimenting to find out what is really true. A cartoon on the internet shows someone interrogating a classic mad scientist. The interrogator asks, "Why did you build a death ray?" The mad scientist says, "To take over the world." "No, I mean what hypothesis are you testing? Are you just making mad observations?" The mad scientist responds, "Look, I'm just trying to take over the world. That's all." The interrogator continues, "You at least are going to have some of the world as a mad control group, right?"

As the cartoon suggests, he's really not a scientist at all; he's just a "mad engineer." (Engineers may understand science, but their goal is not to discover truths about nature but to apply science to make inventions or practical devices.)

Science, Intuition, and Common Sense

Common sense is that which tells us the world is flat.
—Stuart Chase, quoted in S. I. Hayakawa, *Language in Thought and Action*

Common sense is the very antipodes of science.
—Edward Bradford Titchener, *Systematic Psychology: Prolegomena*

We're living in what Carl Sagan correctly termed a demon-haunted world. We have created a Star Wars civilization but we have Paleolithic emotions, medieval institutions and godlike technology. That's dangerous.
—E. O. Wilson, quoted in *New Scientist*

The great biologist (and defender of Darwin) Thomas Henry Huxley wrote, "Science is simply common sense at its best, that is, rigidly accurate in observation, and merciless to fallacy in logic."[10] Scientists and philosophers often claim that science is based on common sense. But at a more fundamental level, much of what we have learned from scientific observations and experiments goes *against* common sense or what we intuitively feel is true.

Think, for example, of how people have viewed the world until just very recently. From our perspective, the sun and moon and stars appear to move around us, and we are the center of everything. From our perspective, the earth looks flat. It takes a lot of early childhood education to train people to perceive the earth as a spherical ball rotating on its axis and revolving around the sun, because that's not what our senses tell us. Our intuition tells us that a heavier or larger object will fall to the ground faster than a smaller or lighter one, and that dogma was carried on from ancient times to the writings of Aristotle and into the Middle Ages. Then Galileo did his famous experiment dropping two different-sized cannonballs off the Leaning Tower of Pisa and showed it was false.

Newton's concept of gravity as attraction between bodies is much less intuitive than the older idea of objects falling to the ground because they had "weight" and everything wanted to move to its "natural place." Even more counterintuitive is thinking about any "solid" object as a collection of tiny nuclei with enormous volumes of space around them, only partially filled with clouds of electromagnetic energy we call electrons. Grasping the enormity of geologic time, with its millions and billions of years, is extremely hard for most people, even with the best analogies and illustrations. Our common sense was evolved when we were small African apes and was not designed to grasp the extremely tiny or the extremely distant.

As Sunil Laxman writes,

This wiring is very deep within us, and starts very early in life. The resistance is not merely limited to viewing some science suspiciously, but for many new ideas that challenge what is apparent. It begins very early in life, with what kids know and learn either by observation and mimicry, or active instruction. Children, even babies, "know" a lot by learning things themselves through observation. They know that solid objects will fall to the ground, for example, or that people have different emotions. Now suppose a child knows that any unsupported object will fall to the ground, it is difficult for this child to imagine or comprehend that the world is round. That is because they have observed that things will always fall off round objects. At a young age, a child cannot comprehend relative scales of the earth (and themselves), and relate it to the concept of gravity. It is just as counter intuitive at that age for a child to believe that a larger object will not fall faster than a smaller object of the same mass, when dropped from the same height. Many of us see that it takes many years for children to be able to accurately draw out the earth as a rounded globe. In essence, people reject scientific ideas because it appears to be counter-intuitive. A level of resistance to science comes from cultural factors. In every culture, some information is specifically asserted or defined. For example, the resistance to understanding evolution is prominent in some parts of America, in certain religious groups. This is because it has been specifically asserted otherwise. Not everyone is qualified to study or understand all scientific principles of a subject (like string theory). Therefore, it's typical for people to believe in what they are told by people they trust. Interestingly, many studies now show that children do the same thing, and will only believe things that are told to them by people they trust. These could be parents, teachers or peers. More importantly, when some data or explanation is contradicting when coming from different sources, children will believe an explanation provided by the people they trust and not the data itself.[11]

The often counterintuitive and difficult-to-grasp nature of science is behind many of the weird ideas about the earth that are discussed in this book. Certainly, flat-earthers and geocentrists are influenced by what they see and intuitively feel, rather than what science tells us. It takes a lot of training to undo natural, "common sense," intuitive perceptions about the world and to grasp the weird, counterintuitive (but correct) views that science has given us.

Baloney Detection

So what are the general principles of science and critical thinking that we need to follow if we wish to separate fact from fiction? How can "deep thoughts . . . be winnowed from deep nonsense?" Many of these were outlined in Carl Sagan's 1996 book, *The Demon-Haunted World*, and Michael Shermer's 1997 book, *Why People Believe Weird Things*. To decipher fact from fiction, some of the most important principles include the following.

1. Extraordinary Claims Require Extraordinary Evidence

This simple statement by Carl Sagan (or the similar "Extraordinary claims require extraordinary proof" by Marcello Truzzi) makes an important point. Every day, science produces hundreds of small hypotheses, which only require a small extension of what is already known to test their validity. But crackpots, fringe scientists, and pseudoscientists are well known for making extraordinary claims about the world and insisting that they are true. These include the many believers in UFOs and aliens, for whom evidence is flimsy at best but who are firmly convinced (as are a majority of Americans, according to polls) that such UFOs have landed here repeatedly and that aliens have interacted with humans. Never mind the fact that such "aliens" seem only to make themselves known to gullible individuals with no other witnesses present or that the "physical evidence" for aliens landing in Area 51 in Nevada or in Roswell, New Mexico, has long ago been explained as caused by secret military experiments. (For further discussion, see *UFOs, Chemtrails and Aliens: What Science Says*, by me and Tim Callahan.)

Just think for a moment: If you were part of a superior alien culture, able to travel between galaxies, would you only interact with a few isolated individuals out in the boonies, or would you contact the head of the governments on this planet and let your existence be known? Think about our extraordinary network of satellites and radar that makes it possible for us to detect virtually anything moving in the skies anywhere in the world. Even with this capability, we have never gotten a reliable detection of a UFO, only unverifiable claims made by random plane or ground observers and photos that have been documented as fakes. Certainly, it is *possible* that aliens have visited us, but such an extraordinary claim requires higher levels of proof than ordinary science, and so far, the evidence provided is pretty flimsy.

As we shall see in this book, most of the weird ideas about the earth are really extreme. They are not obviously false in the way that they are constructed or presented, but in order for us to take them seriously, there must be an extraordinary amount of evidence to support them and to shoot down the evidence of the scientific view. For this reason, most of these ideas are quickly dismissed by real scientists, because there is no evidence for them and lots of evidence against them.

2. Burden of Proof

Related to this first principle is the idea of burden of proof. In a court of law, one side (usually the prosecution or plaintiff) is assigned the task of proving their case "beyond a reasonable doubt" in a criminal case and "based on a preponderance of the evidence" in a civil case. The defense often needs to do nothing if the other side has not met this burden of proof. Similarly, for extraordinary claims that appear to overthrow a large body of knowledge, the burden of proof is also correspondingly greater. In 1859, the idea of evolution was controversial, and the burden of proof was on Darwin to show that evolution had occurred. By now, the evidence for evolution is overwhelming, so the burden of proof on the antievolutionists is much larger; they must show that creationism is right by overwhelming evidence, not point out a few inconsistencies or problems with evolutionary theory. Likewise, the evidence that the Holocaust occurred is overwhelming (many eyewitnesses and victims are still alive, and many Nazi documents describe what they did), so the Holocaust denier has to provide overwhelming evidence to prove that it did not occur.

3. Anecdotes Do Not Make Science

As storytelling animals, humans are prone to believe accounts told by witnesses. Marketers know that if they get a handful of celebrities or sincere-sounding customers to praise their product, we will believe these people and go out and buy their merchandise—even if there have been no careful scientific studies or FDA approvals to back up their claims. One or two anecdotes may sound convincing, and the experience of your back-fence neighbor may be interesting, but to truly evaluate claims made in science (and elsewhere), you need a detailed study with dozens or hundreds of cases. In addition, there often must be a "control" group of individuals who receive a placebo rather than the treatment yet who think that they did get the real medicine, so that the power of suggestion cannot be seen as responsible for the alleged benefit.

Anything approved by the FDA has met this standard; most stuff sold in "New Age" or "health food" stores has not been so carefully studied. When such things have been analyzed, they have usually turned out to have either marginal benefits or none at all. (The stores will take your money all the same.) If you listen closely to the words promoting some of these "medicines," they must carefully avoid the terminology of medicine and pharmacology and must instead use phrases like "supports thyroid health" or "promotes healthy bladder function." These phrases are not true medicinal claims, and so they are not subject to FDA regulations. Nonetheless, the great majority of these products that have been scientifically analyzed turn out to be worthless and a waste of money, and every once in a while, they prove to be harmful or even deadly.

Similarly, the evidence for UFOs or alien abductions or Sasquatch sightings is largely anecdotal. One person, usually alone, is a witness to these extraordinary events and is convinced they are real. However, studies have shown again and again how easily people can hallucinate or be deceived by common natural phenomena into seeing something that really isn't there. A handful of eyewitnesses means nothing in science when the claims are unusual; much more concrete evidence is needed.

4. Arguments from Authority and Credential Mongering

Many people try to win arguments by quoting some authority on the subject in an attempt to intimidate and silence their opponents. Sometimes they are accurately quoting people who really are expert in a subject, but more often than not, the quotation is out of context and does not support their point at all, or the authority is really not that authoritative. This is the usual problem with creationist quotes from authority: when you go back and look at the source, the quote is out of context and means the opposite of what they claim, or the source itself is outdated or not very credible. As Carl Sagan puts it, there are no true authorities; there are people with expertise in certain areas, but nobody is an authority in more than a narrow range of human knowledge.

One of the principal symbols of authority in scholarship and science is the PhD degree. But you don't need a PhD to do good science, and not all people who have science PhDs are good scientists. As those of us who have gone through the ordeal know, a PhD only proves that you can survive a grueling test of endurance in doing research and writing a dissertation on a very narrow topic. It doesn't prove that you are smarter than anyone else or more qualified to render an opinion than anyone else. Because earning a PhD requires enormous focus on one specific area, many people with that degree have actually *lost* a lot of their scholarly breadth and knowledge of other fields in the process of focusing on their thesis.

In particular, it is common for people making extraordinary claims (like creationism or alien abductions or psychic powers) to wear a PhD, if they have one, like a badge, advertise it prominently on their book covers, and feature it in their biographies. They know that it will impress and awe the listener or reader into thinking they are smarter than anyone else or more qualified to pronounce on a topic. Nonsense! Unless the claimant has earned a PhD and done research in the subject being discussed, the degree is entirely irrelevant to the controversy.

For example, many of the critics of the evidence for global climate change are physicists or other scientists with no actual research in climate science. Their degree may make them an expert in physics, but climate science is a completely different field with a different data set and different kind of training. They are presumptuous and arrogant to think that their physics degree makes them an expert in this very different field. Even worse are meteorologists who criticize climate science. Since I teach both subjects at the college level, I can tell you that their claims are ridiculous. Meteorology deals with the day-to-day weather, but climate science deals with climate, the long-term average of weather, based on ice cores, tree rings, deep-sea sediments, and other geological phenomena. A meteorologist has *no* qualification to critique climate data, so when you hear them spouting off about climate change in the news, they come off as rank amateurs. Unfortunately, the average person, who doesn't know that climate is not the same as weather, is fooled nonetheless.

The "scientific" leaders of the creationist movement included a man with a doctorate in hydraulic engineering and another who was a biochemist but trained over seventy years ago. Neither had any training in fossils or in geology or any other field beyond their specializations, but they wrote endless false information about paleontology or geology or thermodynamics. Their doctoral degrees were completely irrelevant to those fields. Yet they always flaunted their PhDs to awe the masses and tried to intimidate their opponents. In all of these cases, a degree in an unrelated field does not make you an expert in any other field. My PhD and published research have made me expert in many areas of geology, paleontology, and climate science, but they don't qualify me to write a symphony or fix a car.

Similarly, there are many fringe ideas in lots of fields, and the more "way out there" they are, the more likely the author has put "PhD" on the cover. By contrast, legitimate scientists do not put their degree on their book cover and seldom list their credentials on a scientific article either. If you doubt this, just look at the science shelves in your local bookstore. The quality of the research must stand by itself, not be propped up by an appeal to authority based on your level of education. To most scientists, credential mongering is a red-flag warning. If the author flaunts a PhD on the cover, beware of the stuff inside!

5. Bold Statements and Scientific-Sounding Language Do Not Make It Science

People who want to promote their radical ideas are prone to exaggeration and famous for making amazing pronouncements such as "a milestone in human history" or "the greatest discovery since Copernicus" or "a revolution in human thinking." Our baloney-detection alarms should go off automatically when we hear politicians or actors hyping policies or movies that turn out to be much less than claimed. The alarms should also scream when we hear people making claims about human knowledge or science that seem overblown.

Another strategy for making a wild idea acceptable to the mainstream is to cloak it in the language of science. This cashes in on the goodwill and credibility that science has in our culture and attempts to make such outrageous ideas more believable. For example, when the creationists realized that they could not pass off their religious beliefs in public school classrooms as science, they began calling themselves "creation-scientists" and eliminating overt references to God in their public school textbooks (but the religious motivation and source of the ideas is still transparently obvious). Several religions (including Christian Science and Scientology) appropriate the aura of scientific authority by using the word in their name, even though the religions are not falsifiable and do not fit the criteria of science as discussed here.

Similarly, the snake oils and nostrums peddled by telemarketers and by New Age alternative-medicine advocates are often described in what appears to be scientific lingo, but when you examine it closely, the makers of the products do not actually follow scientific protocols or the scientific method. We all know examples of television commercials that show an actor in a white lab coat, often with a stethoscope around his or her neck, saying, "I'm not a doctor, but I play one on TV," and then promoting a product. The "doctor" has no medical training, but just the appearance of scientific and medical authority is sufficient to sway people to buy the product.

6. Special Pleading, Moving the Goalposts, and Ad Hoc Hypotheses

In science, when an observation comes up that appears to falsify your hypothesis, it is a good idea to examine the observation closely or to run the experiment again to be sure that it is real. But if the contradictory data are sound, then the original hypothesis is falsified, dead, kaput. It is time to throw it out and come up with a new, possibly better, hypothesis.

In the case of many nonscientific belief systems, from religions to mysticism to Marxism, it does not work this way. Belief systems often have a profound emotional and mystical attachment for people who hold these beliefs in spite of contradictory observations and refuse to let rationality or the facts shake them. As Tertullian put it, "I believe *because* it is incredible."[12] St. Ignatius Loyola, the founder of the Jesuits, wrote, "To be right in everything, we ought always to hold that the white which I see is black, if the Church so decides it."[13] That's fine if you are willing to accept that system and suspend disbelief in favor of emotional and mystical connections.

If you pass off your belief system as science, however, you must play by the rules of science. When con artists try to sell you snake oil and someone points out an inconvenient fact about it, they will try to attack this fact or to explain it away with an after-the-fact or ad hoc (Latin for "for this purpose") explanation. If the snake oil fails to work, they might say, "You didn't use it right" or "It doesn't work on days when the moon is full." If the séance fails to contact the dead, the medium might scold the skeptic by saying, "You didn't believe in it sufficiently" or "The room wasn't dark enough" or "The spirits just don't feel like talking today." If we point out that there are millions of species on earth that could not have fit into the biblical Noah's ark, the creationist tries to salvage their hypothesis by saying, "Only the created kinds were on board" or "Insects and fish don't count" or "God miraculously crammed all these animals into this tiny space, where they lived in harmony for forty days and forty nights" or some similar dodge. Similarly, if you show a claim to be false, the believer may move the goalpost by changing how a falsification of their ideas would be determined.

As we shall see in the chapters that follow, ad hoc hypotheses are common when the conclusion is already accepted and the believer must find any explanation to wiggle out of inconvenient contradictory facts. But they are not acceptable in science. If the conclusion is a given and cannot be rejected or falsified, then it is no longer scientific.

7. Not All "Persecuted Geniuses" Are Right

People trying to promote wild ideas that seem crazy to us will often point to the persecution of Galileo (arrested and tried for advocating Copernican astronomy) or Alfred Wegener (ridiculed for his ideas about continental drift) and will take solace in how these geniuses were eventually proven right. But as Carl Sagan put it, "The fact that some geniuses were laughed at does not imply that all who are laughed at are geniuses. They laughed at Columbus, they laughed at Fulton, they laughed at the Wright Brothers. But they also laughed at Bozo the Clown."[14] The annals of science are full of wild and crackpot notions that didn't survive testing and were eventually abandoned, and these ideas far outnumber the handful of misunderstood geniuses who were vindicated in the end.

These "misunderstood geniuses" often turn to Schopenhauer, who wrote, "All truths pass through three stages. First, it is ridiculed. Second, it is violently opposed. Third, it is accepted as self-evident."[15] But Schopenhauer was wrong. Many revolutionary and radical ideas (such as Einstein's theory of relativity) were never ridiculed or violently opposed. In the case of Einstein, his theories were mostly ignored as interesting but untestable until scientific observations made in 1919 corroborated them.

Science is open to all sorts of ideas, from the conventional to the wacky. It doesn't matter where the ideas come from, but they all have to pass muster. If your ideology has failed the test of science, you can't just claim you're a misunderstood genius; it is more likely that your cherished hypothesis is just plain wrong. Scientists are too busy, and there are too many worthwhile and important scientific goals for them to pursue, for them to waste their time testing and evaluating every wild scheme that comes along. People might wail that they are persecuted and misunderstood geniuses. But if you want to be taken seriously, you must play by the rules of science: get to know other scientists, exchange ideas, be willing to change your own ideas, present your results in scientific conferences, and submit them to the scrutiny of peer-reviewed journals and books. If your ideas can survive this rigorous gauntlet, then they will get the attention they deserve from scientists.

The Skeptic Society in Pasadena, California (I am a member of their editorial board) gets hundreds of letters each year by lone "geniuses" who claim to have made some great discovery, or debunked Einsteinian relativity or quantum physics, or discovered a working perpetual-motion machine or cold fusion or something equally startling. They demand that *Skeptic* magazine publish their "revolutionary" ideas. Most of the ideas are laughably bad and the people clearly crackpots, but every once in a while a somewhat legitimate-sounding idea will emerge, and I am often consulted to see whether it holds muster. But the real test of whether the idea is worthy is peer review. Find a legitimate place to publish your idea, and then let the scientific community test it. If your idea is truly groundbreaking or revolutionary, sooner or later scientists will find its merits and test it, and if it survives repeated scrutiny, scientists will begin to accept it and promote it. Grousing about how you are a misunderstood genius will get you nowhere. Nor will claiming that there is a great conspiracy among scientists to suppress your brilliant idea.

It's a Conspiracy!

Conspiracy thinking, in particular, plays a huge part in weird ideas about the earth. Flat-earthers, geocentrists, moon-landing deniers, creationists, and many others we will discuss in this book insist that they are not taken seriously because a great conspiracy of scientists, or the world in general, is against their ideas. Lately, conspiracy thinking has become rife in society as whole segments of the population are taken in by media that cater to their need for conspiracies, especially shows like Alex Jones's *Infowars*. As the political philosopher John Gray wrote, "Modern political religions may reject Christianity, but they cannot do without demonology. The Jacobins, the Bolsheviks and the Nazis all believed in vast conspiracies against them, as do radical Islamists today. It is never the flaws of human nature that stand in the way of Utopia. It is the workings of evil forces."[16]

Conspiracy theories are everywhere in our culture, and lots of people indulge in them. Ever since President John F. Kennedy (JFK) was shot in 1963, there have been dozens of different conspiracy theories about who shot him and why. Conspiracies have been hatched around Princess Diana's death. Others claim that the moon landing was a hoax (see chap. 6) or that climate change science is a hoax by the entire scientific community trying to destroy capitalism. Just days after the 9/11 terror attacks, a large number of 9/11 "Truthers" emerged to claim that it was all a conspiracy, an inside job by powerful forces—pick your favorite conspirator here—for unstated motives. Of course, the 9/11 attacks *were* a conspiracy—by nineteen Muslim men who hijacked the planes according to a plan hatched by Al-Qaeda. But this is not what the 9/11 Truthers want to accept. It has to be something bigger and more sinister, usually planned by the Bush administration.

High percentages of Americans (on the order of 25–40%) believe in at least one or more conspiracy theory, and studies have shown that those who believe one conspiracy tend to accept many others. Sometimes they are not even consistent. As William Saletan wrote,

> The appeal of these theories—the simplification of complex events to human agency and evil—overrides not just their cumulative implausibility (which, perversely, becomes cumulative plausibility as you buy into the premise) but also, in many cases, their incompatibility. Consider the 2003 survey in which Gallup asked 471 Americans about JFK's death. Thirty-seven percent said the Mafia was involved, 34 percent said the CIA was involved, 18 percent blamed Vice President Johnson, 15 percent blamed the Soviets, and 15 percent blamed the Cubans. If you're doing the math, you've figured out by now that many respondents named more than one culprit. In fact, 21 percent blamed two conspiring groups or individuals, and 12 percent blamed three. The CIA, the Mafia, the Cubans—somehow, they were all in on the plot.[17]

Two years ago, psychologists at the University of Kent led by Michael Wood, who blogs at a delightful website on conspiracy psychology, https://conspiracypsychology.com/author/disinfoagent/, escalated the challenge. They offered UK college students five conspiracy theories about Princess Diana: four in which she was deliberately killed and one in which she faked her death. In a second experiment, they brought up two more theories: Osama Bin Laden was still alive (contrary to reports of his death in a US raid earlier that year), and alternatively, he was already dead before the raid. Sure enough, "The more participants believed that Princess Diana faked her own death, the more they believed that she was murdered," and "the more participants believed that Osama Bin Laden was already dead when U.S. special forces raided his compound in Pakistan, the more they believed he is still alive."[18]

Conspiracy thinking is strongly self-reinforcing. Polls show that those who accepted the JFK assassination conspiracy were twice as likely to believe that a UFO crashed at Roswell (32% believed, versus 16% for those who don't accept any other conspiracy theories).[19] The people who believed in Roswell UFO stories, in turn, were far more likely to believe that the CIA had distributed crack cocaine, that the government "knowingly allowed" the 9/11 attacks, and that the government added fluoride to our water for sinister reasons.

Psychological studies have shown that conspiracy thinking is all about the need for control and certainty in a random, frightening world where everything seems out of control. Conspiracies are nice simple explanations for scary phenomena that we don't want to believe are simply due to random events. Conspiracy believers tend to be people who have high anxiety about their lives, their jobs, and their futures and who need someone to blame for their troubles and failures. Various psychological surveys have shown that believers have a very low level of trust in their fellow human beings or human institutions, tend to have a high degree of political cynicism, and believe the worst about other humans. In broader terms, they are people who focus on intention and agency rather than randomness and complexity.

At one time, conspiracy believers were isolated, and conspiracy thinking was treated as a form of paranoia and mental illness. They had little way of reaching each other, getting feedback from like-minded individuals, or finding lots of new conspiracies to read about and believe in. But now that the internet brings any conspiracy theory to you in the touch of a few keys and mouse clicks, they are proliferating at a rate that has never been seen before, because now they can feed on and reinforce each other. For example, a 2007 poll showed that more than 30 percent of Americans thought that "certain elements in the US government knew the [9/11] attacks were coming but consciously let them proceed for various political, military, and economic motives" or that these government elements "actively planned or assisted some aspects of the attacks."[20] Thanks to relentless conspiracy mongering by the media, polls show that 51 percent of Americans think that a conspiracy was behind Kennedy's assassination; only 25 percent agree with the demonstrated reality that Lee Harvey Oswald acted alone.

One of the worst things about conspiracy theories is the fact they are nearly always airtight; they act like a religion or ideology that refuses to submit to testing and falsification. Every debunking piece of evidence against the conspiracy will be viewed as an attempt to "misinform the public," and the lack of evidence for it is viewed as a government cover-up. In this sense, conspiracy theorists are very antiscientific, because they have the same closed view of the world that will not accept outside information that doesn't fit their core beliefs as religions and cults do. Much about conspiracy groups resembles religious cults, including suspicion of the outside world, self-reinforcement with like-minded individuals, refusal to look at anything that does not fit their worldview, and an almost messianic devotion to the idea that they have the only truth and that everyone else is foolish or deceived or part of the conspiracy.

People have a much easier time believing that a huge operation of sinister forces is at work to do something they don't like rather than accepting the idea that stuff happens. To a conspiracy theorist, the idea that evil forces are ruling the world is much more plausible than the reality that bad things just happen and we don't really have much control over them. Conspiracy thinking is particularly prevalent among people with a deep hatred or distrust of the government, so it tends to be concentrated on the conservative fringe (as evidenced by Donald Trump and his embrace of a wide range of conspiracies and crazy ideas). There is also a strain in conspiracy thinking among leftists who view Big Pharma, Big Tobacco, Big Oil, and so on as more powerful than they really are. We now know that Big Tobacco conspired to suppress antismoking research and that ExxonMobil and some other oil companies conspired to fund climate change deniers and suppress research, but they were not able to hide their conspiracy forever, and the truth came out eventually. As conspiracy thinking also declines slightly with more education, people who know more about how the world actually works tend not to believe in them as much.

So how do we know that the conspiracy believers are wrong and paranoid and that there is nothing really happening? The key flaw with conspiracy thinking is that it assumes a level of competence and secret keeping that has never happened in the history of humanity. People often get the idea from TV and the movies (from shows like *The X-Files* and hundreds of conspiracy-plotted movies, especially spy flicks) that secret government organizations are really powerful and very good at keeping secrets. But the opposite has been demonstrated over and over again. Watergate was a grand conspiracy, but eventually it was exposed. For fifty years, tobacco companies conspired to keep research about the death toll of tobacco under wraps. But whistle-blowers in the companies leaked their top-secret memos, and eventually they were indicted and brought to court and in front of Congress. Leaked documents have shown that ExxonMobil covered up its own climate change research and funded a wide range of front groups and climate-denier groups, using innocent-sounding names to hide their connection to energy companies. Lance Armstrong and just a handful of his closest friends among cyclists knew about his doping activities, but eventually even this tiny circle of silence was broken. And despite the fear of death for breaking the code of silence, or omertà, in the Mafia, sooner or later there is a weak link and the crime bosses go down.

Large secret government operations, like the Bay of Pigs invasion of Cuba, never work as well as they are planned and eventually screw up and are exposed. The Iran-Contra affair was top secret, but eventually a bunch of people made mistakes and it was revealed and investigated. As Michael Shermer quips whenever a 9/11 Truther speaks, "You know how I know it's not a big government conspiracy that's been successfully kept secret for many years? Because it happened during the Bush Administration." Conspiracy believers claimed that the Federal Emergency Management Agency (FEMA) was going to operate concentration camps to keep opponents of the Obama administration under control—which is laughable, because FEMA did not have that capability, as shown by its botched response to Hurricane Katrina in 2005. Also, FEMA employees are not sworn to secrecy. Nearly everything FEMA does is completely open.

Conspiracy theorists claim that these top-secret organizations are capable of hiding everything, but as the Wikileaks and Edward Snowden examples show, sooner or later there is one weak link or leaker who talks or blows the whistle, and government secrets are secret no longer. Donald Trump tried to get away with extorting Ukraine for dirt on his opponent, but a whistleblower exposed that conspiracy and Trump was impeached for it. The Freedom of Information Act has given reporters the power to delve into almost any secret organization, especially governmental organizations, and no secret stays hidden long. The more people and more organizations are required to keep the entire thing hush-hush, the less likely it could actually happen.

On his HBO show *Last Week Tonight*, comedian John Oliver does a hilarious send-up of the entire conspiracy theory mind-set, especially crazy conspiracy YouTube videos. As he puts it, "Conspiracy theories: they're just fairy tales adults tell each other on YouTube." In three minutes, he parodies all the excesses of this way of thinking and "proves" the absurd idea that Cadbury Creme Eggs are a conspiracy by the Illuminati. I highly recommend you watch the video at https://www.youtube.com/watch?v=fNS41ec OaAc, or use your browser to search for it.

The Flat Earth

The "Wisdom" of Celebrities

In 2016 and 2017, the media were abuzz with reports of yet another celebrity suggesting that the earth is flat. Most of the coverage was incredulous and slightly sarcastic, but by giving these ridiculous ideas so much coverage, the media ended up spreading the ideas more widely and even, to some extent, legitimizing them. As tabloid journalism has practiced for years, "if it bleeds, it ledes," and this is even more true now. In today's media world, the whole point of reporting something sensational or crazy, no matter how ridiculous it is, is to get attention and more hits on the website or to sell more magazines. After all, the bottom line is what matters, not the objective truth. But media reports seldom give a critique or a detailed explanation of why 99.99 percent of the world doesn't think the earth is flat.

The media had already created a fuss in 2008 when Sherri Shepherd of the morning talk show *The View* and reality TV personality Tila Tequila said that the earth is flat, or at least questioned the idea that the earth is round.[1] (These same people espoused other discredited notions as well: Shepherd is a creationist, and Tila Tequila has preached a wide variety of controversial ideas, including neo-Nazi antisemitism). A number of prominent professional athletes, including Denver Nugget forward Wilson Chandler,[2] Cleveland Cavalier (now Boston Celtic) guard Kyrie Irving,[3] retired NBA center Shaquille O'Neal,[4] and Minnesota Vikings wide receiver Stefon Diggs,[5] also came out for the flat-earth notion in 2017 and 2018. Irving explained his thinking in the following words:

> Is the world flat or round?—I think you need to do research on it. It's right in front of our faces. I'm telling you it's right in front of our faces. They lie to us. . . . Everything that was put in front of me, I had to be like, "Oh, this is all a facade." Like, this is all something that they ultimately want me to believe in. . . . Question things, but even if an answer doesn't come back,

you're perfectly fine with that, because you were never living in that particular truth. There's a falseness in stories and things that people want you to believe and ultimately what they throw in front of us.[6]

O'Neal is a famous prankster who loves to punk reporters with outrageous statements. He later admitted he was joking just to get a reaction out of people.[7] Irving eventually retracted his statements and gave a public apology to America's science teachers.[8]

The biggest public outrage was the reaction to statements made by rapper B.o.B., whose legal name is Bobby Ray Simmons Jr. B.o.B has advocated the full range of conspiracy theories, including the idea that the moon landing was a hoax, 9/11 was an inside job, the Illuminati are trying to establish a New World Order, Jews are secretly in control of everything, and the US government is actively cloning people. Not only did he start a Twitter war about his beliefs, but he upped the ante, getting into a rap battle with astronomer Neil deGrasse Tyson,[9] repeating all the usual debunked claims of flat-earthers, and even setting up a GoFundMe campaign to raise $200,000 for his own rocket to send up a satellite so that he could see for himself. Like most flat-earthers, he believes that everything from NASA is a hoax, so he wants to do it himself. The idea of sending his own rocket up would be laughable if it were not so sad, and it doesn't consider the problem that even a cheap satellite launch costs about $62 million. Then he recorded and released a rap video called "Flatline," expanding on his ideas, challenging Tyson directly, and even mentioning the noted Holocaust denier David Irving. Some of the lyrics include

> Aye, Neil Tyson need to loosen up his vest.
> They'll probably write that man one hell of a check.
> I see only good things on the horizon.
> That's probably why the horizon is always rising
> Indoctrinated in a cult called science
> And graduated to a club full of liars.[10]

Not to be outdone, Tyson wrote his own rap song, "Flat to Fact," and his nephew, Stephen Tyson, rapped and recorded it. Some of the lyrics include

> Very important that I clear this up.
> You say that Neil's vest is what he needs to loosen up?

The ignorance you're spinning helps to keep people enslaved, I mean
    mentally.
All those strange clouds must be messing with your brain.
I think it's very clear that Bobby didn't read enough
And he's believing all this conspiracy theory stuff.[11]

In March 2018, in a flat-earther stunt, motorcycle racer, daredevil, and
limo driver "Mad" Mike Hughes launched his own homemade rocket almost 1,875 feet into the sky from a homemade launchpad near Amboy, California, on the floor of the Mojave Desert.[12] His intention was to get high
enough to see if the earth really looked curved from space, but at the elevation he reached, it would have been impossible to tell—and he was only
in the sky for less than a minute in a violently vibrating rocket with a tiny
window, after which he made a hard landing and sustained severe injuries.

Hughes told the Associated Press, "I don't believe in science. I know about
aerodynamics and fluid dynamics and how things move through the air, about
the certain size of rocket nozzles, and thrust, but that's not science, that's just
a formula. There's no difference between science and science fiction."[13] Of
course, if he really wanted to see the curvature of the earth from a high altitude, there are lots of safer ways, which are discussed at the end of the chapter,
that would not risk his life and health. On February 22, 2020, Hughes paid
the ultimate price for denying reality when his rocket crashed and killed him.

Hearing all this, most people shake their heads and wonder what has
happened to our society and education system that the weirdest of all ideas
is actively being debated in mainstream media and that someone as famous
as Neil deGrasse Tyson feels it is worth his time to debunk it. Hasn't the
reality of the round earth been established since the time of Columbus? As
Tyson tweeted, "Duude—to be clear: Being five centuries regressed in your
reasoning doesn't mean we all can't still like your music."[14]

As astronomer and author Phil Plait wrote in 2008,

The world is filled with dumbosity, and it's all we can do to fight it. But
sometimes an idea is so ridiculous that you have to wonder if it's a joke.
Yeah, I mean the Flat Earthers. Can people in the 21st century *really* think
the Earth is a flat disk, and not a sphere? When I see their claims I have to
wonder if it's an elaborate hoax, their attempt to poke a hornet's nest just to
see how reality-based people react. The media will sometimes talk to these

goofballs, and I'm glad to report it's almost always tongue-in-cheek, which is probably more than they deserve.[15]

Myths of Columbus

Actually, it's a myth that most people in 1492 thought the earth was flat and that they scorned Columbus because he was convinced it was round. In fact, most educated people have known that the earth is round for at least 2,500 years. Ancient Greeks noticed that the earth cast a curved shadow on the moon during an eclipse. In his dialogue *Timaeus*, Plato wrote that the creator "made the world in the form of a globe, round as from a lathe, having its extremes in every direction equidistant from the centre, the most perfect and the most like itself of all figures."[16] Plato's student Aristotle noticed that if he traveled north or south, it changed which stars he could see above him, and later astronomers discovered that if you traveled to the Southern Hemisphere, the constellations are entirely different.

About 200 BCE, the Hellenistic Greek scholar Eratosthenes famously estimated the circumference and diameter of the earth. He had heard stories that the sunlight shone vertically down into the bottom of a deep well only at high noon on the summer solstice in Syene, about two hundred kilometers south of Alexandria down the Nile River near the modern Aswan High Dam. By using a long rod to measure the length of its shadow and calculating the angle of the vertical rod with the sun overhead in Alexandria, he was able to measure the difference in the angles between Syene and Alexandria (fig. 2.1). Using simple geometry, he estimated the circumference of the earth to be about forty thousand kilometers. This is amazingly accurate, less than 0.16 percent off the value that we now accept.

Nor did the discoveries of the Greeks die with the "Dark Ages" and the loss of most of the writings of the ancient Greeks and Romans. Although some medieval scholars thought the earth was flat, most of them had read Plato and Aristotle and accepted their evidence that the earth was round. About 1250 CE, the medieval scholar John Sacrobosco wrote *Treatise on a Sphere*, with multiple proofs of the curvature of the earth. In it, he said,

> That the earth, too, is round is shown thus. The signs and stars do not rise and set the same for all men everywhere but rise and set sooner for those in the east than for those in the west; and of this there is no other cause than

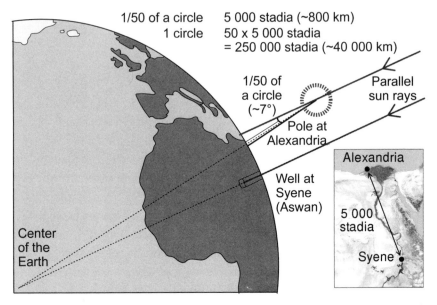

Figure 2.1. Diagram showing Eratosthenes's famous experiment to calculate the size and curvature of the earth. He noticed that at summer solstice, the sun was directly overhead in Syene (which is on the Tropic of Cancer, where the sun's rays come straight down on the first day of summer). Meanwhile, to the north where he lived in Alexandria, the sun made a 7° angle from a post sticking vertically above the ground. Eratosthenes used this angle and the known distance between Syene and Alexandria to calculate the size of the earth to within 1 percent of the values we know now. (*Courtesy Wikimedia Commons.*)

the bulge of the earth. Moreover, celestial phenomena evidence that they rise sooner for Orientals than for westerners. For one and the same eclipse of the moon which appears to us in the first hour of the night appears to Orientals about the third hour of the night, which proves that they had night and sunset before we did, of which setting the bulge of the earth is the cause.[17]

The myth that most educated people in the medieval times and up to 1492 believed in a flat earth is a relatively recent notion. As Jeffrey Burton Russell documented in his 1991 book, *Inventing the Flat Earth: Columbus and Modern Historians*, American author Washington Irving, famous for his stories of Rip van Winkle and the Headless Horseman of "The Legend of Sleepy Hollow," created this fiction; he needed to spice up the conflict between the Church and Columbus in order to improve the drama for his

1828 book, *A History of the Life and Voyages of Christopher Columbus*. Irving was very widely read and cited, so his myth entered all the American history textbooks for the next century. Even as late as 1983, it was still widely believed, and the myth appeared in historian Daniel Boorstin's best-selling book, *The Discoverers*.

Modern Flat-Earthism

In fact, flat-earth beliefs were a rare fringe idea with few followers until relatively recently. In the 1800s, the most famous flat-earther was Samuel Rowbotham (1816–1884). In the 1860s, he pioneered the modern flat-earther notion that the earth was a disk centered over the North Pole (fig. 2.2), bounded on its outer edge by a wall of ice (instead of Antarctica over the South Pole, which cannot exist in their version of geography). The skies above were a dome of fixed stars only five thousand kilometers above the earth's surface, consistent with the old medieval notion of the heavens before the birth of modern astronomy. His ideas were first published in a pamphlet called *Zetetic Astronomy*, followed by a book called *Earth Is Not a Globe*, and another pamphlet, *The Inconsistency of Modern Astronomy and Its Opposition to the Scriptures*, which revealed the biblical literalist roots of most flat-earth thinking.

According to Rowbotham, the "Bible, alongside our senses, supported the idea that the earth was flat and immovable and this essential truth should not be set aside for a system based solely on human conjecture."[18] He is correct in saying this, because there are at least sixteen places where the Bible says the earth is flat; talks about the "four corners of the earth," the "ends of the earth," and the "circle of the earth"; or suggests that you can see the entire earth from a high place.[19] Rowbotham and later followers like William Carpenter and Lady Elizabeth Blount kept promoting the idea and founded the Universal Zetetic Society after Rowbotham's death in 1884. This incarnation of flat-earth thinking died out some time after 1904.

After about fifty years of virtually no organized activity, the rebirth of flat-earth thinking occurred in 1956 with the founding of Samuel Shenton's International Flat Earth Research Society, based in his home in Dover, England. Always a tiny group, with a very limited membership, they corresponded through a homemade mailed newsletter, yet every once in a while, they managed to get a short burst of publicity in the newspapers. In the 1960s and 1970s, when Mercury, Gemini, and Apollo astronauts first began

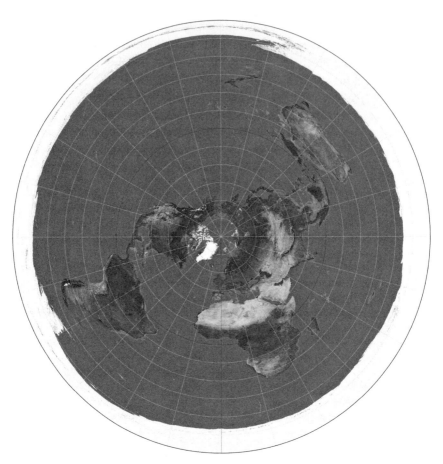

Figure 2.2. Map of the earth from a north polar projection. (*Courtesy NASA.*)

to produce images of the earth from space, Shenton dismissed the images as hoaxes (the common belief among flat-earthers ever since), saying, "It's easy to see how a photograph like that could fool the untrained eye."[20] Later, he attributed the curvature of the earth seen in NASA photographs to a trick of the curvature of wide-angle lenses: "It's a deception of the public and it isn't right."[21]

After Shenton's death in 1971, Charles K. Johnson picked up the mantle and inherited Shenton's library from his wife. He reorganized the group as the International Flat Earth Research Society of America and Covenant People's Church, where they maintained their lonely quest at his home in the town of Lancaster in the Mojave Desert.[22] They claimed to have reached a membership as large as 3,500, scattered around the world, paying annual dues of six to ten dollars. The society communicated via the quarterly *Flat Earth News*, a four-page tabloid written and edited almost entirely by Johnson and sent in the mail. As hard-core biblical literalists, they emphasized all the passages that state that the earth is flat. Every few years, they would get smirking coverage in the newspapers, but their membership declined during the 1990s, especially after a fire at Johnson's house in 1997 destroyed all records and membership contact information. Johnson's wife died shortly afterward, and then the society itself vanished when Johnson died on March 19, 2001.

Flat-earth thinking might still be a tiny fringe belief with no organized leadership were it not for the internet and the ability of believers all around the earth to find each other and organize a virtual community. In 2004, the Flat Earth Society was resurrected by Daniel Shenton (no relation to Samuel) as a web-based discussion forum and then eventually relaunched as an official society, with a large web presence and their own wiki.[23] As of July 2017, they claimed a membership of five hundred people. However, the publicity from celebrity entertainers and musicians, such as those discussed at the beginning of this chapter, seems to suggest that flat-earth ideas are much more common (see chap. 18), even if the believers are not official members of the Flat Earth Society. There are a number of other flat-earth societies on the internet not affiliated to Shenton's group. The first Flat Earth International Conference met in Raleigh, North Carolina, on November 9 and 10, 2017, with about five hundred attendees.[24] In May 2018, there was a three-day flat-earth convention in Birmingham, England, with several hundred attendees who traveled all the way to England to hear a spectrum of speakers with a common belief in the flat earth.[25] Even more alarming, about a third of millennials are not convinced that the earth is round (as discussed in chap. 18).[26] And there are calls on the internet for a reality show to let the flat-earthers test their ideas and actually try to travel off the edge of the earth![27]

In 2018, Netflix produced a documentary about the flat-earthers called *Behind the Curve*.[28] Like most such documentaries, it consists mostly of interviews of the major advocates of a particular idea (in this case, the flat earth) and contrasting views of other interviewees who regard the believers as crazy. It starts with one of the stars of the flat-earth movement, Mark Sargent, a middle-aged, balding man who still lives with his mother and depends on her to feed him. Sargent spouts one incredible claim after another, sitting in his mother's basement obsessing over little details and posting hundreds of YouTube videos expounding his ideas. He claims as proof of the flat earth that he can see skyscrapers from his mother's Whidbey Island backyard. However, Whidbey Island is less than forty-eight kilometers (about thirty miles) from downtown Seattle, too close to detect the curvature.

Sargent describes how he obsessed for three solid days trying to track aircraft online that flew near or across the South Pole and then decided there weren't any such flights. According to Sargent, this proves that Antarctica is not a continent on the South Pole but a giant ice wall on the perimeter of the flat earth. (Later in the same part of the movie, a Caltech grad student pulls up a different flight tracking site and finds plenty of planes flying over parts of Antarctica.) He shows his handmade model of the flat earth with the dome of the sky and stars above it, and the moon and sun rotating in the sky above us, but he does not explain how this would create the phases of the moon or would explain eclipses, which are entirely impossible with his model.

When you argue with a flat-earther, a highly revealing moment is when they fall back on their cop-out "Oh, that's just math and physics—I don't believe in those." In the documentary, Sargent says, "The reason why we're winning against science is that science just throws math at us," as if that were some mark of how smart he is and how he is beating science. This is behind much of their thinking: they are only capable of simple intuitive models and are typically math-phobic, so they refuse to do even the simplest calculations that would show why their ideas are impossible. By contrast, since the days of Isaac Newton, the reasons we know the earth is round are best understood by doing mathematical calculations that only make sense in a spherical globe and cannot be accommodated in a flat earth.

But the most revealing moment in the documentary is when the flat-earthers attempt to do experiments to prove their point. In both cases, the experiments actually show that the earth is round, and the flat-earthers refuse to accept the results:

One of the more jaw-dropping segments of the documentary comes when Bob Knodel, one of the hosts on a popular Flat Earth YouTube channel, walks viewers through an experiment involving a laser gyroscope. As the Earth rotates, the gyroscope appears to lean off-axis, staying in its original position as the Earth's curvature changes in relation. "What we found is, is when we turned on that gyroscope we found that we were picking up a drift. A 15 degree per hour drift," Knodel says, acknowledging that the gyroscope's behavior confirmed to exactly what you'd expect from a gyroscope on a rotating globe. "Now, obviously we were taken aback by that. 'Wow, that's kind of a problem,'" Knodel says. "We obviously were not willing to accept that, and so we started looking for ways to disprove it was actually registering the motion of the Earth." Despite further experimental refinements, Knodel's gyroscope consistently behaves as if the Earth is round. Yet Knodel's beliefs seem unchanged when discussing the experiment at a Flat Earth meetup in Denver. "We don't want to blow this, you know? When you've got $20,000 in this freaking gyro. If we dumped what we found right now, it would be bad. It would be bad. What I just told you was confidential," Knodel says to another Flat Earther in attendance.[29]

The second experiment was run by Knodel's cohost on his flat-earth YouTube channel, Jeran Campanella. This experiment provides the ending for the film. As described in *Newsweek*,

Campanella devises an experiment involving three posts of the same height and a high-powered laser. The idea is to set up three measuring posts over a nearly 4 mile length of equal elevation. Once the laser is activated at the first post, its height can be measured at the other two. If the laser is at eight feet on the first post, then five feet at the second, then it indicates the measuring posts are set upon the Earth's curvature.

In his first attempt, Campanella's laser light spread out too much over the distance, making an accurate measurement impossible. But at the very end of *Behind the Curve*, Campanella comes up with a similar experiment, this time involving a light instead of a laser. With two holes cut into styrofoam sheets at the same height, Campanella hopes to demonstrate that a light shone through the first hole will appear on a camera behind the second hole, indicating that a light, set at the same height as the holes, travelled straight across the surface of the Flat Earth. But if the light needs to be raised to a different height than the holes, it would indicate a curvature, invalidating the Flat Earth.

Campanella watches when the light is activated at the same height as the holes, but the light can't be seen on the camera screen. "Lift up your light, way above your head," Campanella says. With the compensation made for the curvature of the Earth, the light immediately appears on the camera. "Interesting," Campanella says. "That's interesting." The documentary ends.[30]

Even more revealing than the failure of their experiments and their reactions when they inadvertently demonstrate the curvature of the earth is the insight into the psychology of flat-earthers. Like many other conspiracy believers and cult followers, flat-eartherism is a fundamental belief system to them and a community, so flat-earthers cannot allow anything to change their minds. Otherwise, they will lose their sense of identity and group belonging as well as their feeling of understanding and controlling the world around them. As reported in *Newsweek*,

> "Say you lose faith in this thing. What then happens to my personal relationships? And what's the benefit for me doing that? Will the mainstream people welcome me back? No, they couldn't care less. But, have I now lost all of my friends in this community? Yes. So, suddenly, you're doubly isolated," psychologist Dr. Per Espen Stoknes says in the documentary. "It becomes a question of identity. Who am I in this world? And I can define myself through this struggle." "If I tried to go. . ." [flat-earther Mark] Sargent says in the documentary, contemplating the scenario described by Dr. Stoknes. "They would come and say, 'Don't, don't do it.' So I couldn't, even if I wanted to."[31]

How Do We Know?

One thing we have learned from this widespread skepticism of science and established reality is that we scientists and educators need to do a better job of conveying both the facts of science and the *evidence* of those facts to people. We need to describe and demonstrate the evidence *why* we know certain things to be true. As Neil deGrasse Tyson wrote, "The fact that there's a rise of Flat-Earthers is evidence of two things. One, we live in a country that protects free speech. And two, we live in a country with a failed educational system. . . . Our system needs to train you not only what to know, but how to think about information and knowledge and evidence. If we don't have that kind of training, you'd run around believing anything."[32]

So how do we know that the earth is roughly spherical in shape?[33] How could you tell for yourself without engaging in dangerous stunts like launching yourself in a homemade rocket? To answer these questions, we will not use observations from satellites, spacecraft, or aircraft, because flat-earthers believe that these are all hoaxes and part of a giant conspiracy.

1. Watch ships at sea: Even before the Greeks wrote about the spherical earth, ancient seafarers knew that if you watch a ship sail away to the horizon, the bottom hull of the ship vanishes first, followed by the mast and then the top of the ship (fig. 2.3). If it is sailing toward you, you see the masts first, followed by the hull as it gets closer. This only makes sense if the ship is sailing around the curve of the earth. Flat-earthers also have heard of this evidence, of course, and claim it is an illusion caused by the perspective on different objects. But this is not how perspective works. If an object is far away on a flat surface, it will get smaller, but the lower part will not vanish as it recedes; instead, all of it will get smaller but remain fully in view. This is true even if you go to a harbor and follow the ship using a telescope or binoculars to improve your distance vision. The ship will vanish from bottom to top, not just become smaller.

Figure 2.3. Medieval drawing of a ship at sea disappearing bottom first on the horizon of a curved earth. (*Public domain.*)

2. Look to the stars: As the ancients noticed, the constellations look different as you travel north and south in latitude on the earth. About 350 BCE, Aristotle was one of the first to record this observation. Traveling from Greece to Egypt, he could see the difference in the skies. As he noted, "There are stars seen in Egypt which are not seen in northern regions." He realized that the earth was small enough that its curvature was apparent over that relatively short distance, "for otherwise the effect of so slight a change of place would not be quickly apparent."[34] The difference became even more obvious when the first European explorers traveled south of the equator and found a whole new sky full of unfamiliar stars and constellations. The Southern Cross, for example, cannot be seen until you travel south of the Florida Keys, yet it becomes the major constellation of the sky when you are south of the equator. Meanwhile, the Big Dipper, which dominates the night sky above forty-one degrees north latitude, vanishes below the horizon as you head south, so at about twenty-five degrees south latitude in northern Australia, it is gone from the sky.

3. **Watch a lunar eclipse:** Every few years, we experience a lunar eclipse, where the disk of the full moon is covered by the shadow of the earth. It's weird to watch the circle of the moon gradually get darker and darker as the edge of the earth's shadow gradually covers it (fig. 2.4). As first discovered by the ancients and reported by Aristotle, the edge of earth's shadow is unmistakably curved and becomes even more so as the eclipse approaches totality. Finally, the earth's shadow covers the moon completely, so the only moonlight you see is from light that has passed around the curve of the earth and through our atmosphere (turning it red), refracting to the middle of the shadow.

Many times, the lunar eclipse is not total, but as the distance of the earth from the moon increases, it casts a slightly smaller shadow and the shining edges of the moon are visible on the edge of the shadow. This is called an annular eclipse, and it shows the entire shadow of the earth as a circle or ring of light around the dark shadow. This would never make sense if the earth were flat. Flat-earthers claim that the sunlight is blocked by the flat circular disk of the earth, but why then does the sun never happen to catch the flat disk of the earth on its edge or at an angle, so the shadow has a shape other than a circle? The only way this is possible is if eclipses happened only at midnight, when the "flat disk" is perpendicular to the sun-earth axis, so the "dark side" of the earth would only see the total eclipse of the moon when it was directly overhead at the stroke of midnight. In fact, lunar eclipses happen at all different times of day and night (although they are not very visible in the daytime).

Figure 2.4. A total lunar eclipse on June 15, 2011, as seen from Budapest, Hungary. The upper left frame shows totality, with the moon entirely covered by the earth's shadow. Over the next hour, the earth's shadow moves off to the lower right, and its distinctly curved edge can be seen, showing that the earth casts a curved shadow and therefore must be a spherical shape. By 23:10 Universal Time, the shadow has almost completely vanished and the full moon is visible. (*Photo courtesy Wikimedia Commons.*)

4. Go climb a mountain: If the earth were flat, you could see huge distances if you had a good enough telescope, so looking across the distance between Miami and New York City, only 1,000 miles or 1,760 kilometers, should be no problem. But if you are standing on level ground, even under the best of conditions with a superpowerful telescope, you can see no farther than about 3 miles (5 kilometers). Any object farther than that disappears below the horizon. Of course, if you climb a tree or even a mountain, you can see a bit farther on a day with excellent clear air and visibility. Standing on a hill 60 meters high, you can see about 50 kilometers. But even from the tallest mountains, no one can see much farther than about 60 miles or 100 kilometers, certainly not the distance from New York to Miami.

Take another example: Mauna Kea volcano on the Big Island of Hawaii is the highest peak in the Hawaiian Islands at 4,205 meters (13,796 feet). On a flat earth with nothing but ocean for many miles, you should be able to see enormous distances on a clear day. On the island of Kauai, only 487 kilometers (303 miles) away, is that island's highest peak, Kawaikini, at 1,592 meters (5,226 feet). Over such a short distance on a flat earth, someone on Mauna Kea should easily be able to see the top of Kauai, but you can't, because the earth is curved. Thanks to that curvature, the farthest you can see from Mauna Kea is 374 kilometers (233 miles).

5. Go fly in a plane: If you fly around the world, you are traveling around a sphere. You cannot do this on a disk-shaped earth. If you calculated the distances to travel in a circle around the North Pole on a disk-shaped earth (fig. 2.2), it would not add up to the distance you must actually travel around a spherical globe, no matter which latitude you traveled along. Even more convincing is the view from high above the earth. Unlike the people like "Mad" Mike Hughes who killed himself in his homemade rocket that rose only 1,875 feet, there are ways to get high enough to see the earth's curvature. In a passenger jet flying above 35,000 feet, the curvature begins to be visible, although you need a wide window with a sixty-degree field of view to detect the curvature.

This isn't possible to see with the tiny passenger windows, but the crew on the flight deck can see it fine, so anyone in the cockpit in flight can see it. (Sadly, after the 9/11 hijackings, the flight deck is always locked against intruders during flight.) Above fifty thousand feet, the curvature becomes more obvious, although few commercial aircraft fly that high. The now-retired supersonic Concorde jet routinely cruised at sixty thousand feet, so passengers on those flights could see the curvature of the earth easily. And of course, military aircraft and spacecraft and our thousands of satellites fly much higher and see it all the time, but as flat-earthers believe that everything from NASA and the military is part of great conspiracy to hoax us all, that won't help convince them.

6. Fly near the South Pole: Flat-earthers claim that there is no South Pole or Antarctic continent over it, but just a 1,500-foot-tall ice wall around the edge of the earth's disk guarded by NASA (fig. 2.2). According to the Flat Earth Society, no one has been past this ice wall and lived to tell the tale. Of course, this makes everyone who has ever traveled to

Antarctica a hoaxer and liar, including pioneering polar explorers like Roald Amundsen (who reached the South Pole first), Ernest Shackleton, Sir Robert Scott, and others who made these expeditions before the flat-earth idea of the ice wall had been suggested. It also makes liars out of anyone else who may have traveled across the Antarctic Circle and returned successfully, or all the polar researchers down in Antarctica right now. (I have several friends down there finding fossils as I write this.)

Despite what flat-earthers claim, commercial flights do travel over part of the Antarctic,[35] and if you get the right window seat and good weather during these flights, you can see parts of Antarctica from your seat. Most commercial flights across the Southern Hemisphere don't fly over the center of Antarctica because it is not on the shortest possible route (the great circle route, or the straightest line on a globe) between South America and Australia, or South Africa and Australia. But they do fly over the edge of the continent, so you could look down and see the Antarctic ice sheet from your window seat.[36] A flight between New Zealand and South Africa would cross Antarctica, but currently there are no flights scheduled to do this.[37] Anyway, a commercial flight over the ice cap is not a good idea, especially given the bad weather over Antarctica most of the year—and also because if they have plane trouble, it's much better to make an emergency landing in the Southern Ocean where there is a chance of rescue rather than in the middle of the Antarctic ice cap. Flights that run south of seventy-two degrees south latitude must carry special survival gear in case they go down in the polar region. As this regulation reduces the number of paying passengers they carry,[38] not many flights are scheduled in the Antarctic Circle.

7. **Send up a balloon:** Another way to get your own images from high enough to see the earth's curvature is to send up a weather balloon. Both balloons and the kinds of cameras and equipment needed to record and transmit the signals are easily available through commercial sources now, for anyone who has the technical skills and funds to try this. In January 2017, a group of students from the University of Leicester Department of Physics of Astronomy and members of the Leicester Astronomy and Rocketry Society did just such an experiment. Their weather balloon, launched from Tewksbury in Gloucestershire, rose 77,429 feet (23.6 kilometers) into the sky, and their cameras sent back stunning footage of the curved earth from the high atmosphere (fig. 2.5).[39] You are welcome

Figure 2.5. Image of earth from balloon. (*Courtesy Wikimedia Commons.*)

to watch the footage online for yourself (just search for videos under "Project Aether") and see it vividly demonstrating the view from higher and higher elevations. After reaching its maximum altitude (where the temperature was about–56° Celsius and the air pressure was nearly a vacuum), the payload then descended to earth at speeds of more than 100 miles per hour and was successfully recovered in Warwickshire. You can try it yourself if you have the money and the expertise! Just contact the Federal Aviation Administration before you launch to make sure that your balloon doesn't fly into restricted airspace.

8. **Compare shadows:** If you are motivated, you can replicate Eratosthenes's famous experiment (fig. 2.1) yourself. The simplest way to do it would be to take a long flight in a north-south direction. Before you take off, measure the length your shadow casts at a particular time of day at your starting point. Take the flight, and then at the same time on the following day, measure the shadow at your new location. It should be measurably longer or shorter if you have traveled far enough. If the earth were a flat disk and not a sphere, this would not happen, because the sunlight coming in at an angle on a flat disk would always cast the same length of shadow.

9. Compare time zones: As anyone with jet lag can tell you, traveling east to west, or west to east, around the earth for any significant distance is discombobulating, because changing time zones upsets your biological clock. This is a direct demonstration of how different parts of the spinning earth are facing the sun at different angles, so they are all experiencing a different time of the day relative to the sun. The easiest way to confirm this in this world of instantaneous satellite communication (which in itself is a confirmation of the spherical earth) is to compare the time you are experiencing in your area with the time of someone in a different part of the world. For example, if it's noon in New York and you email or text or call a friend in Beijing, it's midnight there, and it's 1:30 a.m. in Adelaide, Australia. Or if you look up the times for sunrise and sunset at different longitudes around the earth, you see that they occur at different times. This is simply impossible with a flat earth. Flat-earthers have tried to get around this problem by claiming that the sun's light casts a big circular flat "spotlight" that is pointing in a circle around the different parts of the earth, but that explanation falls apart if you think about it. If you were in a large darkened theater, you could still see the spotlight casting its light on the stage even though you might be in total darkness where you sit.

10. Compare seasons: If the earth were flat, the sun's rays would hit all parts of the earth from straight above and would not come in at an angle, like they actually do. In addition, there would be no seasons, because on a flat earth, both the Northern and Southern Hemispheres would get the same amount of solar radiation all year round. We would all experience whatever seasons there were the same way. But thanks to the spherical shape of the earth and its tilted axis, we experience seasons at different times, so winter in the Northern Hemisphere is summer in the Southern Hemisphere, and vice versa.

11. Feel the pull of gravity: If you accept the laws of gravity worked out by Isaac Newton, then the force of attraction of gravity should get stronger the closer you are to the center of mass. In a flat-earth disk (fig. 2.2), gravity should be strongest in the center of the disk at the North Pole and much weaker as you approach the Antarctic, on the edge of the disk. If you dropped an apple in Australia or southern Patagonia, it should fall somewhat sideways, pulled toward the North Pole, not straight down—but it doesn't. Anyone with a decent gravimeter (a device that measures gravitational attraction) can measure the gravity anywhere on earth, and the attraction at sea level is nearly always the same (except for local effects like crustal rock beneath you, which is denser or less dense than average). The scientific data for these gravity measurements have been gathered for more than a century and are widely published, although flat-earthers reject all scientists as part of the global conspiracy that includes NASA. (Of course, many flat-earthers don't believe in gravity, either, but fall back on Aristotle's antiquated notions that objects fall because they are heavier or lighter.)

12. **Consider the solar system:** Anyone with a decent telescope, good night visibility, and a chance to look at the moon or the other planets night after night can confirm what early astronomers (especially Galileo) could see: all the other bodies in the solar system are spherical. You can see the shape of the moon clearly, especially as the shape of the illuminated part of the moon changes with the cycle of the full and new moons every month. After several nights of closely observing Jupiter in a good telescope, you can confirm that it is round and spinning on its axis; even better, each night you can see its four largest moons moving around as they orbit around it. It's a lot more difficult to do this with Mars or Saturn, but Galileo was able to see it when he first trained a telescope to the skies. So if all the other bodies in space are spherical, why would only the earth be flat?

If that's not convincing enough to fair-minded flat-earthers or to people sitting on the fence on the issue or having doubts about the shape of the earth, then they are hopelessly lost in the mind-set of a cultist, and no amount of evidence will convince them. Sometimes they accidentally reveal the disconnect between their worldview and reality, as when a tweet from the Flat Earth Society read, "The Flat Earth Society has members around the globe." Then there is the humorous internet meme "The only thing flat-earthers have to fear is sphere itself." So we will consider this case closed and will move on to the next weird idea about earth: geocentrism.

3

Ptolemy Revisited

Religion versus Science

The scene is a dramatic one, depicted in many plays and novels. Galileo Galilei, now sixty-eight years old, gray haired, losing his sight, and bowed with age, stands before the Inquisition in early 1633 (fig. 3.1). He is on trial for heresy, since he has advocated the notion that the sun is the center of the solar system (heliocentrism), a view first published by Nicholas Copernicus in 1543. This contrasts with geocentrism, the plain, intuitive common-sense idea that the earth, not the sun, is the center of the universe, as Aristotle and people since earliest times long believed (fig. 3.2). In 1609, using one of the first telescopes ever built (invented almost simultaneously in Holland and in Venice) to study the night sky, Galileo concluded that the earth moved around the sun, and he published his celestial observations and ideas in a short book titled *Sidereus Nuncius* ("Heavenly Messenger") in 1610.

By 1616, the Inquisition was scrutinizing his works for signs of heresy. They warned Galileo that he could talk about heliocentrism in a theoretical way but not actually claim that the earth really did move around the sun. Trying to steer clear of antagonizing the Church, Galileo argued that heliocentrism could be made consistent with religious dogma. But Church leaders disagreed, citing such passages as Psalms 93:1 and 96:10 and many other verses. In 1 Chronicles 16:30, the Bible says, "The world is firmly established, it cannot be moved." Psalm 105:5 reads, "The Lord has set the earth on its foundations; it can never be moved." Another example is Ecclesiastes 1:5, which says, "And the sun rises and sets and returns to its place." And in Joshua 10:12, the Hebrew leader Joshua calls on the Lord to make the sun stand still so that they can continue their battle. Thus, Galileo was fighting not only thousands of years of established belief in geocentrism but also the Scriptures, since Europe was still ruled by either the Catholic Church or the Protestants.

Figure 3.1. Painting of Galileo defiantly facing the Inquisition. (*Courtesy Wikimedia Commons.*)

The last straw for the Church fathers was the 1632 publication of Galileo's provocative book *Dialogue Concerning the Two Chief World Systems*. Encouraged by the election of his supporter Cardinal Maffeo Barbarini as Pope Urban VIII and the support of some Italian nobles, Galileo framed the argument as a dialogue, a miniature play between three characters. This allowed him to make his argument by letting the characters debate the ideas among themselves, as Plato had done with the philosophical debates of Socrates. Galileo could put his own ideas in the words of one character (Salviati, the scientist in the dialogue) but could plausibly claim that he himself wasn't advocating heliocentrism; the heretical ideas were spoken by only one character in his "play."

Salviati presents the Copernican arguments for heliocentrism in a debate with the other main character, Simplicio, who represents the traditional geocentric views of Ptolemy, Aristotle, and the Church. Although the name

Un missionnaire du moyen âge raconte qu'il avait trouvé le point
où le ciel et la Terre se touchent...

Figure 3.2. Famous 1888 engraving by Flammarion showing the old geocentric view of the universe. The image depicts a man crawling under the edge of the sky, depicted as if it were a solid hemisphere, to look at the mysterious Empyrean beyond. The caption translates to "A medieval missionary tells that he has found the point where heaven and Earth meet." (*Courtesy Wikimedia Commons.*)

was drawn from Simplicius of Cilicia, a sixth-century commenter on Aristotle, it was a deliberate double entendre, since *Simplicio* also meant "simpleton" or "fool" in Italian. The third character in the play is Sagredo, an intelligent but uncommitted layman and merchant who acts as the target audience and jury for the arguments of the other two. He eventually agrees with Salviati's heliocentric solar system—as would any reader, since Salviati demolishes Simplicio's arguments.

Galileo thought that he had cleverly avoided being accused of directly advocating heliocentrism as a true description of the world, but the Inquisition was not amused. The pope did not like the public ridicule that was clear in the character of Simplicio. Galileo was called to Rome to face several days of trial, where they showed him instruments of torture and made him recant his views on pain of torture and death. As he bowed before them and confessed his rejection of the heretical heliocentric views, Galileo supposedly muttered under his breath, "*Eppur si muove*" ("And yet it moves!"). Rather than torture the old man, the pope sentenced him to house arrest, where he spent the last ten years of his life unable to leave his domain. There, Galileo wrote his last great work, *Discourse and Mathematical Demonstrations Relating to Two New Sciences*, where he laid down the foundations of modern physics, especially regarding the motions of objects (kinematics).

By 1638, Galileo was completely blind (partially due to staring directly at the sun with his crude telescope to see the sunspots) and was suffering from a painful hernia and insomnia, and his devoted daughter had to take care of him. He died on January 8, 1642, at the age of seventy-seven. His *Dialogues*, along with Copernicus's work, was banned by the Church, so they could not be printed or read except in places where the Church's power was limited. Galileo's books and his ideas were still under official Church ban until 1835, even as the rest of the world had moved on to modern astronomy thanks to the work of Isaac Newton in the early 1700s. Finally, on Halloween 1992, Pope John Paul II officially acknowledged the errors of the Catholic tribunals and announced that a statue of Galileo would be placed in the Vatican. In December 2008, Pope Benedict XVI praised Galileo's work on the four hundredth anniversary of Galileo's earliest telescopic observations. However, the plan to put his statue in the Vatican has since been shelved.

The idea of a heliocentric solar system is most famously attributed to the Polish scholar Mikolaja Kopenika (in Polish; we know him by the Latinized version of his name, Nicolaus Copernicus), although a version of the idea was first proposed by the Greek scholar Aristarchos of Samos about 280 BCE. Copernicus was a true genius; fluent in Latin, German, and Polish, and also speaking some Greek, Italian, and Hebrew, he often worked as a translator. He is more famous as an astronomer and mathematician, although he also served as a diplomat and governor. As an economist, he formulated the quantity theory of money, an idea that later came to be known as Gresham's law.

But Copernicus is most famous for making the astronomical observations that led him to suggest heliocentrism and give strong evidence for it.

In the years 1512–1515, he made a long series of intensive observations of stars and planets. Among the puzzles that he dealt with was a curious phenomenon known to the ancient Greek astronomers as *retrograde motion*. If you studied the position of certain planets, like Mars and Jupiter, against the background of the "fixed stars" night after night, you would observe something odd. Each night, the planets seem to have moved farther in the sky than the previous night, as if they were circling the earth. But once in a while, the planet appeared to pause, then back up a short distance, before resuming its former forward motion. This backward, or "retrograde," motion (fig. 3.3A) made no sense if planets simply orbited the earth in a simple circle (or, later, an ellipse).

Many solutions to this puzzle were proposed, but the most famous was by the Hellenistic Greek astronomer Claudius Ptolemaeus (known as Ptolemy today) who lived in Alexandria around 100–170 CE, during the days of the early Roman Empire. He viewed the universe as a set of nested spheres, all spinning around the earth at the center (fig. 3.2). Each of the planets was on a sphere spinning around the earth, and the "dome of the sky" covered with the "fixed stars" was the outermost shell of the spheres. His *Almagest* summarized nearly all the known observations of the stars and planets up to that time, so it was the foundation of all later astronomy.

To explain the odd backward motion of certain planets, he postulated that they didn't move in a single circular path; instead, they were moving around a small circle (epicycle) whose center was the larger circle of their motion around the earth (fig. 3.3B). Sometimes during their motion around the earth, they would be on the reverse-moving part of the epicycle, so they would appear to move backward from the perspective of the earth. This idea was soon the most popular among all the astronomers. By the time the Church dominated all Western thought, they made Ptolemy's system the officially approved model of the universe, just as Aristotle's ideas about nature were also considered officially sanctioned by the Church. For over 1,400 years, no one dared challenge the Ptolemaic system.

Copernicus was dissatisfied with Ptolemy's explanation of retrograde motion but not because he was a rebel who wanted to challenge the Church or the dogma of his day. Instead, he disliked Ptolemy's system of epicycles because it seemed too complicated and inelegant. He sought a simpler explanation that made sense without all the tinkering and fudging that astronomers had to do to make the epicycles work. Eventually, Copernicus realized

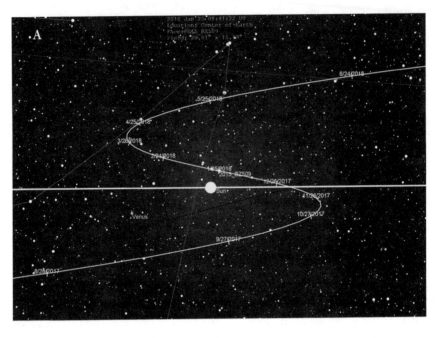

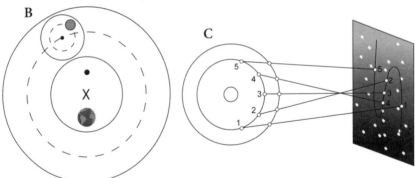

Figure 3.3. Retrograde motion. A. The apparent motion of asteroid 514107 2015 BZ509 against the background of fixed stars as it is viewed night after night. B. Ptolemy's explanation of retrograde motion, where planets move in epicycles centered around a point in orbit around the earth. C. Copernicus's explanation, where apparent retrograde motion is caused when the faster-moving earth (numbers 1–5) on a short inner track catches up and passes a slower-moving outer planet like Mars or Jupiter (dots on outer track). The projection on the right shows how the outer planet would be seen from earth. It appears to backtrack as the earth passes it on the inside track. (*Courtesy Wikimedia Commons.*)

that if the sun, rather than earth, was the center of the system, it all made sense. If Mars and Jupiter were on orbits around the sun but outside earth's orbit, then they would be moving in much bigger circles and much more slowly than us.

Let's imagine that Mars or Jupiter is ahead of us in its orbit (fig. 3.3C) and the earth comes up fast behind them, around its much shorter inside orbit, until it passes Mars or Jupiter. From the earthbound perspective, Mars will appear to move forward then appear to back up as we overtake it on the inside bend. After we pass it completely, it will appear to move forward again. It's analogous to two race cars going around a big curve. The car on the inside of the bend has less distance to travel, so it will often pull ahead of a car leading it on the outside bend. From the inside driver's perspective, the outside car appears to slow down and fall back as it is passed, even though from the spectators' view, all the cars are moving forward. We travelers on the race car that is earth are on a faster-moving vehicle on the shorter inside bend from outer planets like Mars and Jupiter, so each time we come up from behind and pass them, they briefly appear to be falling back.

This simple but elegant solution first came to Copernicus shortly after his observations in 1515, but reluctant to publish, he spent a lot of time writing it up and tinkering with it. Some of his students wrote and published short summaries of his ideas, so they were known to many scholars and some Church officials. Copernicus himself wasn't in any hurry until late in his life, when he finally wrote it all down as *De Revolutionibus Orbium Coelestium* ("On the Revolution of Heavenly Spheres"). Copernicus was justifiably afraid of criticism from the Ptolemaic astronomers of the time and especially from the Church—so much so that he dedicated the work to Pope Paul III in hopes of placating religious authorities. The book finally went to press in 1543, just as Copernicus was dying of a stroke and paralysis at age seventy. Legend has it that he was shown the final printed pages of his book just before he died, knowing that his work would be published.

Once he died, Copernicus was safe from the firestorm of anger, criticism, and censure that his work provoked and from the torture of the Inquisition. After his death, his work was mostly ignored as purely theoretical for decades, and not until people like Giordano Bruno (burned at the stake for his heretical views) and Galileo revived it did the Church consider his work a threat and ban Copernicus's book. As we just discussed, it was not until the 1990s that the Church finally made official peace with Copernicus and Galileo, even though their work had become the foundation of modern astronomy with the work of Newton in the early 1700s.

"Galileo Was Wrong: The Church Was Right"

Most readers of this book might be thinking, "OK, so much for the history of how the heliocentric solar system was discovered. After all, scientists proved it in the 1700s, and even the Catholic Church finally recanted, only 350 years late." That's what I thought too, until I was startled to find mention of a seminar held at Notre Dame University in South Bend, Indiana, on November 6, 2010, titled "Galileo Was Wrong: The Church Was Right." At first I thought it might be some kind of clever satire, but once you clicked on the web page (since taken down), it was clear that they were dead serious! Who were these people, and how is it that they have a significant following in the twenty-first century?

Robert Steinback of the Southern Poverty Law Center, which monitors racists and antisemites, attended the meeting and described it as follows:

> "Seminar" might be generous phrasing: The presentation was a mind-numbing, 15-hour-long sermon-cum-pep rally for radical traditionalist Catholic apologists desperate to debunk any science that suggests the Bible shouldn't be interpreted literally. Numerous biblical passages describe the earth as at rest, with the sun in transit around it. About 90 mostly Catholic devotees, curious skeptics and feisty college students who converged on South Bend endured a series of one-sided monologues declaring that theorists from Copernicus and Galileo to Einstein and Hawking were wrong about celestial physics. Though much of it was difficult for mathematical mortals to follow, the presenters' gambit was clear enough: Can anyone really prove the earth isn't sitting still? That's tougher than it sounds: Even though astrophysicists tell us that every body in the universe is in motion, it will always appear that the thing you're on is standing still relative to everything else.
>
> "If we're saying that the earth is in the center of the universe and it's not moving, that means Someone, with a capital S, put it there," said the conference's principal speaker and emcee, Robert Sungenis, founder of Catholic Apologetics International. . . .
>
> Some of the South Bend presenters have taken their certainty about what they see as literal biblical truth to hateful extremes far more

consequential than dismissing Galileo: Sungenis has published a number of venomously anti-Semitic screeds that drew official church condemnation and have rendered him unwelcome in most mainstream Catholic circles.

E. Michael Jones, who also was at the gathering, has used his South Bend-based magazine *Culture Wars* to viciously denounce the "Jewish world view" and has expressed enthusiasm for many core Nazi ideas about Jews (a sampling of his magazine's cover stories: "Judaizing: Then and Now," "The Judaism of Hitler" and "Shylock Comes to Notre Dame"). Martin G. Selbrede, who also spoke, is vice-president of the Chalcedon Foundation, the leading think tank of the Bible-literalist Christian Reconstruction theology; the foundation has never renounced the racist, homophobic and anti-Semitic views of its late founder, R. J. Rushdoony. . . .

The best moments of the gathering came when 15 or so bright college students finally got to confront the purported experts at the end of the long day. The panel struggled with and occasionally mocked their questions and host Sungenis at times seemed to bristle at the students' audacity.

If such an attitude is typical of all like-minded theorists, it's easy to scientifically postulate that geocentrism is a lot of solar hot air.[1]

The summary of the seminar also captures the main features of the modern geocentrist movement. They're a tiny group of extreme Catholics (sometimes called *traditionalist Catholics*) who reject most of the changes in their Church in the past few decades, including the pope's apologies to Galileo and acceptance of modern science. As the article also states, they fall back to an even earlier phase of Church history, when antisemitism and persecution of Jews was one of their main habits (many Jews were tortured and executed by the same Inquisition that tried Galileo).

Their leader is a man named Robert Sungenis, who got a bachelor's degree in religion from George Washington University and a master's degree in theology from Westminster Theological Seminary. He calls himself "Dr." Sungenis because he got a "doctorate" from an unaccredited online diploma mill that calls itself "Calamus International University," incorporated in the Republic of Vanuatu. He began in a Catholic family, then converted to Protestantism as a young man, and then swung back to the extreme form of Catholicism in his later years. Sungenis attributes his conversion to geocentrism to reading the book *Geocentricity* by the creationist Gerardus Bouw in 2002. (Ironically, most of the modern creationists who are literalist about every other part of the Bible reject geocentrism.)

By 2006, he had become a major advocate of geocentrism; had self-published (with Robert Bennett) a three-volume book, *Galileo Was Wrong: The Church Was Right*; and was running a website, www.galileowaswrong.com.[2] The site is full of slick video clips, blog posts, and a forum promoting their ideas. One video audaciously claims that geocentrism is "the coming scientific revolution." Just like creationist websites, it is full of attacks on scientists and scientific ideas, wild claims with only minimal fact-checking from anyone outside their community, and a distinct sense of paranoia that the entire world is against them because they are on to the truth.

Sungenis's organization has done their share of stunts to please their following and thumb their nose at the rest of the world. In 2006, they offered a $1,000 reward to anyone who could prove that the earth moves around the sun. Like most such contests by pseudoscientists, the conditions of the award are so restrictive that no one can satisfy them, and those who have successfully demonstrated heliocentrism have been rebuffed on one technicality or another.[3]

Their most outrageous stunt was a 2014 ambush film called *The Principle*. Using a different tentative film title and concealing their true motivations, Sungenis and executive producer Rick DeLano obtained interviews with numerous distinguished scientists, including Lawrence Krauss, Michio Kaku, Max Tegmark, Julian Barbour, and George F. R. Ellis, and they paid actress Kate Mulgrew (best known for her appearances on *Star Trek: Voyager*) to narrate it. They asked the right kinds of questions to make these scientists think that it was a genuine, honest documentary, let their guards down, and say things that could be edited to emphasize the uncertainty of science. These included questions about controversial topics like dark matter and multiverses as well as segments edited to sound like the scientists support geocentrism.

The physicists who were ambushed responded in no uncertain terms, according to Colin Lecher in an article in *Popular Science*:

> Along with Krauss, at least two of the mainstream scientists who appear in the film aren't so happy about it. Max Tegmark, a brilliant MIT cosmologist and science communicator, is spoken of admiringly by DeLano in the radio show. When I asked about his appearance in the film, Tegmark

emailed: "They cleverly tricked a whole bunch of us scientists into thinking that they were independent filmmakers doing an ordinary cosmology documentary, without mentioning anything about their hidden agenda or that people like Sungenis were involved." Ditto for South African mathematician and cosmologist George Ellis, a well-respected professor at the University of Cape Town who wrote *The Large Scale Structure of Space-Time* with Stephen Hawking. "I was interviewed for it but they did not disclose this agenda, which of course is nonsense," he wrote me. "I don't think it's worth responding to—it just gives them publicity. To ignore is the best policy. But for the record, I totally disavow that silly agenda."[4]

This film's deceptive tactics and the ambush interviews mirror the similar efforts of the "intelligent design" creationists in their 2008 film *Expelled: No Intelligence Allowed*. Hosted by the obscure character actor and right-wing celebrity Ben Stein, it ambushed a number of distinguished scientists and skeptics (including my friends Michael Shermer, head of the Skeptic Society, and Eugenie Scott, then director of the National Center for Science Education) with questions that sounded odd to them at the time. Their responses were then edited to sound like there was a giant conspiracy to suppress intelligent-design creationism from discussion in the public arena. *Expelled* opened to universal bad reviews (except in the evangelical circles, where it was required viewing) and made so little money that eventually its production company went bankrupt in 2011. But it created a lot of fuss before it flopped, which was the whole point.

The same might be said of *The Principle*. It screened in only a few theaters starting on October 24, 2014, and as of 2015, it had grossed a measly $89,543, much less than it cost to make.[5] But the film was expected to lose money; it had a different goal. As Colin Lecher explained,

Despite its absurdity, the mere fact that DeLano, Sungenis, and the rest of their crew were able to fund and execute a slickly produced film, and to cajole famous physicists to sit and chat for it, makes the geocentrist fringe startlingly real: people who believe in these ideas not only exist, but have the wherewithal to make a movie. There's nothing simple about producing a film, much less one with some of the most technically-minded people on the planet. In DeLano's case, he is (or at least was) apparently steadily employed, eventually on chummy terms with a respected production company, and seems intimately familiar with science, even though his interpretations of it are a minority view, to put it charitably. If the film is absurd (it surely is), its creation was something clear-eyed, thought through.

Why did the creators bother to make the film if they realized that the respected scientists appearing would immediately denounce it? There's the chance they didn't expect the denouncements, but that seems unlikely. Another possibility, suggested by DeLano's initial eagerness to talk to me, was that the establishment backlash had been part of the plan all along. Surely *The Principle*, after those countless media reports—including this one—is in a better position than it was before, even if potential viewers check it out only for novelty's sake. Even if it's fleeting, being the center of the universe has its perks.[6]

As they say in show business, there's no such thing as bad publicity. Anything that gets you noticed, no matter how critical or negative, gets you attention you might not otherwise have—and that was the whole point. There's no sign that the film changed a lot of minds (especially since it was barely seen by anyone) or that geocentrism is a growing movement. For example, there have been no more repeats of the 2010 geocentrism conference, while there are now annual flat-earther meetings, and many organizations that tout creationism meet around the calendar.

Why do these people care so strongly about an issue that was settled over 350 years ago? The answer, as they say in so many of their documents and interviews, is religion. To them, anything that takes humans out of the center of the universe makes humans insignificant and no longer the center of God's creation. Indeed, that was the reason for much of the resistance to heliocentrism in the early days. The Church was not only wedded to literal interpretations of Scripture; it also felt that humans were the apple of God's eye and could not possibly be living anywhere but in the center of God's creation. Many other people have noticed this too. For example, in 1917, Sigmund Freud wrote,

In the course of centuries the naïve self-love of men has had to submit to two major blows at the hands of science. The first was when they learnt that our earth was not the center of the universe but only a tiny fragment of a cosmic system of scarcely imaginable vastness. This is associated in our minds with the name of Copernicus, though something similar had already been asserted by Alexandrian science. The second blow fell when biological research destroyed man's supposedly privileged place in creation and proved his descent from the animal kingdom and his ineradicable animal nature. This revaluation has been accomplished in our own days by Darwin, Wallace and their predecessors, though not without the most violent contemporary opposition.[7]

Indeed, we need no more evidence of this than the words of Sungenis himself: "You can't have the earth at the center of the universe by chance.... The devil is a powerful foe and he will use something like [the model of a sun-centered solar system] to win his battle. If [scientists] have to admit that the earth is in the center of the universe, where does the power shift back to? It shifts back to the church."[8]

How Do We Know?

As we have already mentioned in chapter 1, the common sense, intuitive view that humans have held since prehistoric times is that the sun, moon, and planets *appear* to be moving around us; therefore, the earth is the center of the universe. For us to visualize the system differently and think of ourselves as moving around the sun requires an early education that violates our senses and intuition. Many ideas in science do not agree with common sense; they are nonintuitive and require imagination and a lot of training to understand and accept. Yet that is what the evidence, from Copernicus to today, demonstrates.

As we did in chapter 2, it is worthwhile to briefly describe some of the evidence and observations that support the heliocentric model and falsify geocentrism. Science may not always be easy to understand, but as its methods and results are constantly challenged, tested, and subjected to peer review, they stand the test of time. It is important for any educated human in the twenty-first century to know some of this evidence, so that people understand why science supports heliocentrism. You should not just accept it because you were told to believe it while you were in school.

First, how do we know that the earth is rotating on its axis and that the sunrise and sunset are not caused by the sun going around us but by the earth rotating with respect to the sun?

1. **Watch it from space:** Obviously, the most straightforward evidence comes from spacecraft, which have repeatedly photographed the

movement of the earth in real time. For example, there are videos showing the earth rotating as viewed by the Galileo spacecraft,[9] and you can locate many by just searching for "earth rotation Galileo" in your browser. But modern geocentrists are much like flat-earthers and regard all evidence from NASA and all the other international space agencies as part of a big global government conspiracy involving all the world's astronomers and space scientists, so that won't convince one of them. This also applies to the extraterrestrial space telescopes like Hubble and Gaia, which are in orbit around the sun, not orbiting the earth, so they can see the earth's motion from outside our sphere of influence.

2. **Foucault's pendulum:** If you have been to some of the modern science museums or public observatories (like Griffith Park Planetarium in Los Angeles or Hayden Planetarium in New York), you might have seen a room where a long pendulum (fig. 3.4) is suspended from a high ceiling; its base is a big circular platform with a series of pegs or other markers around the edges. (There are many good video demonstrations online if you search for "Foucault pendulum.") If you watch the pendulum for a while, you will see it slowly knock down one peg after another.

 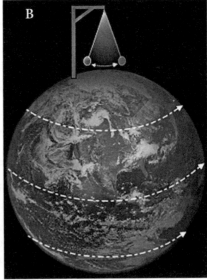

Figure 3.4. A. Foucault's pendulum. (Photo by the author.) B. Diagram showing how Foucault pendulum over the North Pole would swing in one plane but on earth would appear to move in a circle. (*Courtesy Wikimedia Commons.*)

When you first see it, if you note which peg has been knocked over, go spend some time looking at other exhibits, and return before you leave. You will probably see that the pendulum has knocked down another peg. (More modern versions have lights that are triggered when the pendulum passes over them.)

This experiment is known as Foucault's pendulum, first demonstrated by French physicist Léon Foucault in 1851. If you set the pendulum in motion on a motionless earth, it would continue to swing back and forth in a single plane, as long as enough energy is supplied to keep the pendulum from slowing down and stopping. But the earth is rotating beneath the pendulum, so as soon as the pendulum starts, the earth moves a certain number of degrees beneath it every hour. This makes the pendulum appear to move around in a circle, but what is really happening is that the pendulum is moving in the same plane and the earth is turning beneath it. Modern geocentrists have no good explanation for this except vague references to Mach's principle, which concerns the difficulty of describing anything in an absolute reference frame.

3. Coriolis effect: Another even larger result of the earth's rotation is the Coriolis effect, something I have to explain early in every class I teach about oceanography, meteorology, and climate change, since it is fundamental to the way oceanic and atmospheric currents move around the globe. You can demonstrate it on a kid's playground merry-go-round (fig. 3.5). If you are on one side of the spinning merry-go-round and try

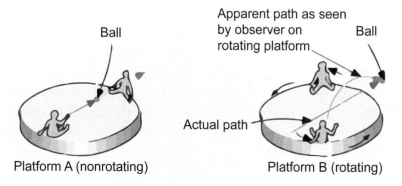

Figure 3.5. The Coriolis effect. On a merry-go-round, if you throw a ball at a target across the spinning disk, it will miss, since the target is moving away from the spot to where you threw the ball. (*Courtesy Wikimedia Commons.*)

to throw a ball to your friend on the opposite side, the ball will appear to curve away from your friend (to the right if it's spinning counterclockwise; to the left if it's spinning clockwise). In simplest terms, this is because your friend is a moving target, so as soon as you release the ball thrown straight at him, he moves away from the point you targeted where he used to be, and the ball will miss him. It appears to curve sideways from your rotating perspective, but it's actually moving in a straight path; you and your friend are doing the actual moving. (There are several excellent videos demonstrating this online, if you just type "Coriolis" into your browser.)

The same goes for the motion of the currents of air and water around the world. If the world were not spinning, the air would rise from the tropics (where there is an excess of solar heat and the warm ground is heating the air, constantly creating a plume of rising air and low pressure), then move due north and south from the equator to the poles, where it would descend in a permanent zone of high pressure on the poles. But thanks to Coriolis, the air in the Northern Hemisphere curves to the right as it moves, creating the great circulating belts of air in different latitudes known as the Hadley, Ferrel, and Polar cells as well as permanent features like the west-going subtropical trade winds and the east-moving prevailing westerly winds in the middle latitudes.

These same winds drive the surface ocean currents of the world, creating the enormous circuits of water in the tropics and subtropics known as gyres, which move in a giant counterclockwise loop in the Northern Hemisphere and a clockwise loop in the Southern Hemisphere. And the huge cyclonic storms, such as hurricanes and typhoons, always rotate counterclockwise in the Northern Hemisphere and clockwise in the Southern Hemisphere—all due to Coriolis. It even works on smaller scales, such as a giant long-distance cannon. A sniper shooting in the middle latitudes of the Northern Hemisphere would find the shot deflected 7 centimeters (3 inches) to the right if he or she shoots 1,000 meters (about 3,300 feet). Modern geocentrists have no explanation for this global phenomenon.

4. **The Chandler wobble:** The earth's rotation is not perfectly smooth. Instead, the earth wobbles on its axis very slightly over long periods of time, a phenomenon known as the Chandler wobble. It makes the stars and galaxies visible in the sky appear to wobble around their normal

positions if you observe them over thousands of years. If the earth were stationary, then the modern geocentrist would have to explain why all the stars and galaxies wobble in the same direction and the exact same amount. Furthermore, measurements show that some stars and galaxies are relatively close to us (say, five light-years away), while others are farther (say, ten light-years away). Since the light we see from them started at different times (five years ago versus ten years ago), for them all to wobble the same amount would require enormous coordination and synchronicity, which violates all the laws of physics.

5. **Motions of other planets:** All the other planets in our solar system are spinning on their axes, something that can easily be observed with a good telescope that can resolve the surface features of Jupiter. If they are all spinning on their axis, why is the earth the only body that is not rotating?

What about proof that the earth is revolving around the sun? Again, the phenomenon is so large in scale that it's difficult for us to see on earth, but it does provide a number of successful predictions that can be observed and tested.

1. **Observations from space:** As pointed out before, both modern geocentrists and flat-earthers reject images from space as hoaxes perpetrated by the great conspiracy of NASA, other international space agencies, and the entire worldwide scientific community. Nevertheless, the Hubble and Gaia space telescopes have repeatedly shot images of the earth in different parts of its path around the sun. Even more impressive are recent Mars rover images of a sunrise over the surface of Mars, something that would not happen in the same way if Mars were in an orbit around the earth and the sun were on an inner orbital track.

2. **Phases of Venus:** As Galileo first observed and published in 1610, Venus has phases (full Venus, half Venus, three-quarters Venus) just like our moon, something that could not happen if both the sun and Venus were orbiting the earth. It makes sense only if Venus is orbiting the sun (fig. 3.6). To get around this problem, modern geocentrists adopt a weird hybrid system proposed by Tycho Brahe as a compromise between geocentrism and heliocentrism in which the sun orbits the earth but the rest of the planets orbit the sun.

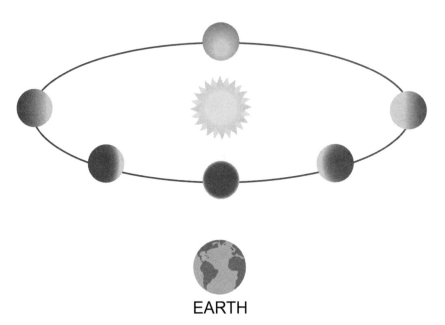

EARTH

Figure 3.6. Diagram showing the phases of Venus, only explicable if the sun is at the center of the orbits of Venus and the earth. (*Courtesy Wikimedia Commons.*)

3. Retrograde motion: As Copernicus pointed out, the phenomenon of retrograde motion requires extremely complicated and unlikely gyrations, such as Ptolemy's epicycles (fig. 3.2), to work in a geocentric system but is much more simply explained by heliocentrism. Once again, modern geocentrists fall back on Tycho's weird hybrid system to explain it.

4. Stellar parallax: For a long time, early astronomers rejected the idea of the earth's motion around the sun because of the lack of apparent stellar parallax. They reasoned that if the earth traveled in a huge ellipse around the sun (186 million miles or 300,000 kilometers in diameter), the position of the closer stars against the background of the most distant stars should be slightly different when looked at on one side of the orbit and again on the opposite side of the orbit six months later. Since the astronomers could not detect any difference in the stars, they initially rejected the heliocentric model. It turns out that there *is* a parallax effect, but most of the stars are so much farther away from us than the early astronomers thought that it is hard for us to perceive on earth. Finally, in 1838, astronomer Friedrich Wilhelm Bessel success-fully demonstrated that there is a parallax effect in the stars. It takes extraordinarily careful measurements to detect it, since most stars are so far away from the earth that the parallax effect is tiny.

5. **Starlight aberration:** Imagine that you are standing still and the rain is coming straight down on top of you. If you hold your umbrella straight up, it will shield you from drops descending vertically. But if you are moving into the rainstorm, you need to tilt your umbrella "into" the rainstorm to keep protected, even though the rain is still coming straight down. The faster you move, the more you must tilt your umbrella. Likewise, if the earth were not revolving around the sun, starlight would come straight down on us, but if it is moving, then there would be a tilt effect of the starlight coming in at an angle. English astronomer James Bradley first detected the aberration of starlight in 1825, ironically while he was making measurements to demonstrate stellar parallax (but failing at it).

6. **The speed of Neptune:** Neptune is so far away from us (four light-hours, or the distance it takes light to travel in four hours; about 2.6 billion miles) and on such a long orbital path that it would have to travel faster than the speed of light to circle the earth in just twenty-four hours, as it appears to do. This is physically impossible, but geocentrists will still make misguided appeals to Einstein's theory of relativity to dodge the problem. It is not a problem, however, if Neptune is slowly revolving around the sun, which is what it is actually doing.

There is no need to belabor the point any further. The evidence for a spinning earth revolving around the sun is overwhelming to any fair-mind-ed person, and only the religious extremism of the modern geocentrists makes them twist scientific data into incredible knots in order to preserve ideas that have been discredited for over four hundred years.

The Hollow Earth

Myths of the Underworld

Many ancient cultures believed that there was an underworld of some sort beneath us, whether it was the hell of the Christian tradition (She'ol to the Hebrews), the underworld caverns of the god Hades of the ancient Greeks, Cruachan to the Celts of Ireland, Patala of the Hindu tradition, Svartálfaheimr of the Nordic mythology, or Shamballa in the Tibetan Buddhist legend, among many others. In some cases, the underworld was a place where people went after death, while in other cultures, the ancestors of living people emerged from the underworld. In his legendary poem *Inferno*, Dante Alighieri wrote about entering the Gates of Hell:

> Through me you pass into the city of woe:
> Through me you pass into eternal pain:
> Through me among the people lost for aye.
> Justice the founder of my fabric moved:
> To rear me was the task of power divine,
> Supremest wisdom, and primeval love.
> Before me things create were none, save things
> Eternal, and eternal I shall endure.
> All hope abandon, ye who enter here![1]

These legends were apparently inspired by the mystery of large caves that could not be fully explored or that descended into darkness beyond the people's spelunking ability. Sometimes these were mysterious mountain caves, while others appear to be based on the drained conduits from volcanoes called lava tubes. Caves have always held a fascination and a terrifying mystery for many cultures, especially when their early explorers found the bones of dead people or animals in them. Indeed, the mystery of caves is a common theme in literature, from Plato's cave allegory, to Dante's *Inferno*, to Tom Sawyer, Becky Thatcher, and Injun Joe getting lost in a cave in the climax of Mark Twain's novel *Tom Sawyer*.

From these legends, it is not surprising that most people even in the early scientific literature imagined that the mysterious caves of the earth led to a great hollow world beneath. The famous astronomer Edmond Halley (best known for correctly figuring out the path of the comet now named for him) also had some questionable ideas. He thought that the earth was a hollow shell about 500 kilometers (500 miles) thick, with two additional inner shells surrounding a solid metal core. From this geometry, he imagined that these shells had their own magnetic fields and that the aurora borealis was formed by gases escaping from these hollows inside the earth. Science fiction and fantasy writers L. Sprague de Camp and Willi Ley claimed that the great mathematician Leonhard Euler thought that the hollow earth had its own sun inside. Sir John Leslie imagined there were two central suns inside the earth. This last idea inspired Jules Verne when he wrote one of the first works of science fiction, *Journey to the Center of the Earth*, in 1864.

In the nineteenth century, there were many more theories about what the inside of the earth was like. One of the most famous and influential, by the War of 1812 veteran John Cleves Symmes Jr., postulated that the earth is a hollow shell about 1,300 kilometers (810 miles) thick, with openings at both poles that were about 2,300 kilometers (1,400 miles) across. Inside this were four additional inner shells, each with openings at the poles. Symmes wrote, "I declare that the Earth is hollow and habitable within; containing a number of solid concentric spheres, one within the other, and that it is open at the poles 12 or 16 degrees."[2] He pledged his life to promoting his notion, boldly declaring, "I am ready to explore the hollow." Symmes then spent years touring the United States with a handmade wooden globe that opened to show the spherical inner layers in an attempt to raise money and get the US government interested in an expedition to one of the poles to find these openings.

In 1822, Senator Richard Thompson petitioned Congress to supply Symmes with "the equipment of two vessels of 250 to 300 tons for the expedition, and the granting of such other aid as Government may deem requisite."[3] It was debated in Congress for a while, and seven bills were introduced in the House, but nothing was passed. After Symmes's death, one of his followers, Jeremiah Reynolds, continued to preach the need to find these polar openings and even went to Antarctica himself. Such agitation led to the

US Exploring Expedition of 1838–1842. That expedition spent four years sailing the Pacific and tried to sail to the South Pole but only got as far as Disappointment Bay in Antarctica before being forced to return. Later followers, such as James McBride, Jeremiah Reynolds, Professor W. F. Lyons, and Symmes's son Americus, all published books advocating and amplifying Symmes's theories about the hollow earth. In 1885, William Fairfield Warren wrote *Paradise Found: The Cradle of the Human Race at the North Pole*, arguing that humans came from inside the earth through a hole in the Arctic and that the Eskimos and Mongols of the polar regions were their direct descendants.

The idea was common in fiction as well. Even before Verne's famous 1864 novel, in 1838, Edgar Allan Poe wrote *The Narrative of Arthur Gordon Pym of Nantucket*, a pioneering fictional account of a journey inside the earth. In 1892, science fiction author William R. Bradshaw wrote a novel, *The Goddess of Atvatabar*, claiming that the earth's interior was filled by a world he called Atvatabar (fig. 4.1).

Through the early twentieth century, numerous books and articles proposed different versions of the hollow earth, some as works of science fiction but others claiming to be works of real science. Tarzan creator Edgar Rice Burroughs wrote *At the Earth's Core* in 1914. By the 1950s and 1960s, most of the hollow-earth ideas were also linked to ideas that Bigfoot or aliens came from the interior of the earth and emerged from caves, and many others tied their notions of the hollow earth to the legends of Atlantis and flying saucers.

As wilder and wilder notions were propagated in the twentieth century, actual scientific research was also taking place to understand what really was going on in the earth's interior. The notions of the hollow earth have not subsided in the least, despite all the scientific evidence discussed at the end of this chapter. Instead, they have moved into the realm of the fantasy and the paranormal, where their proponents, oblivious to scientific evidence, are convinced that there is a giant NASA conspiracy to hide the holes in the poles leading to the underworld (a common theme among many of these fringe notions such as flat-earthers and geocentrists and others discussed elsewhere in the book). Many of these ideas will come up if you just do a simple internet search for "hollow earth." Instead of dissecting them all, let's look at a couple in detail to get the flavor of how these people think and what they claim.

Figure 4.1. A cross-sectional drawing of the earth showing the "Interior World" of Atvatabar, from William R. Bradshaw's 1892 science-fiction novel *The Goddess of Atvatabar.* (*Courtesy Wikimedia Commons.*)

Peter Kolosimo, a follower of Erich Von Däniken's ancient astronaut theories, asserted that a robot was seen entering a subterranean tunnel below a monastery in Mongolia and that a light from the underground was seen in a cave in Azerbaijan. Kolosimo and other advocates of ancient astronaut ideas, like Robert Charroux, think that UFOs and aliens live inside the earth. A similar idea was pushed by "Dr. Raymond Bernard" in his 1964 book, *The Hollow Earth.* He claimed that flying saucers and UFOs emerge from the earth's interior and that Atlantis is connected to the earth's interior.[4] To support his idea, he suggested that the hollow earth is like the Ring Nebula in space, which is a gigantic mass of stars and cosmic dust, enormous compared to the earth. Then, skeptical investigator Martin Gardner did some research and found that "Dr. Raymond Bernard" is a pseudonym for another fringe author, Walter Kafton-Minkel. As recently as 1989, the title of Walter Kafton-Minkel's book *Subterranean Worlds: 100,000 Years of Dragons, Dwarfs, the Dead, Lost Races and UFOs from Inside the Earth* indicates how far into the realm of paranormal the entire notion of the hollow earth had gone.

A number of pseudohistoric accounts are common among modern hollow-earthers. One is the supposed account of Karl Unger, a German sailor on a 1943 U-boat expedition to the South Pole, which allegedly entered the hollow earth through an underwater passageway. There they reached a place called Rainbow Island, where an advanced civilization welcomed them. In fact, Hitler was famous for believing all sorts of legendary and paranormal nonsense, and he apparently believed in the hollow earth as well. Some conspiracy theorists claimed that Hitler did not die in his bunker in 1945 but escaped into the hollow earth and is still living there.

Another piece of fake history concerns Admiral Richard Byrd, the first man to fly over the poles, first to the North Pole in 1926, then across the Atlantic in 1927, and to the South Pole in 1929. There is supposedly a "hushed-up" secret diary of a 1947 flight to the North Pole, where he described a land full of lush greenery and lakes and woolly mammoths living inside the earth. One major problem with the story: Byrd was actually flying over Antarctica at that time.

Will Storr, author of *The Unpersuadables: Adventures with the Enemies of Science*, wrote about some of the stranger hollow-earthers. One of these is Rodney M. Cluff, author of *World Top Secret: Our Earth Is Hollow!* As Storr describes Cluff's account,

> "I was working on a New Mexico farm when I was 16 and the farm manager's son started talking about it," he says. Fascinated, he began reading up. "I found evidence from the Scriptures, history and science that our Earth is hollow as well as all the planets and the moons and even asteroids."
>
> So convinced was Cluff that, in 1981, he flew his wife and five children from New Mexico to a new life in Alaska. "I thought, 'Why don't we see if we can find the way to the Hollow Earth?'" And was his wife keen? "She wanted to go back home. She thought I was crazy. But we did it anyway."
>
> In Alaska, Cluff met a small group of people who had travelled to the icy state with the same idea. Soon they were ready to embark upon their mission. "We started on the road up to Point Barrow," he says. "We saw a sign, at one point, saying 'This Is A Private Road: Don't Go Any Further.' So we didn't go any further."
>
> How long did he drive before he reached the sign and aborted the mission? "About an hour," he says. There's a silence while I process this information. "Less than. . ." Soon afterwards, the Cluff family moved back south.
>
> Cluff, for his part, tried to reach the Hollow Earth again. In 2003, he received an email from a man named Steve Currey who'd recently inherited his family's travel firm that specialised in far-flung expeditions. Currey

had once heard his father talking about the Hollow Earth and was familiar with Cluff's book. They decided to plan a new trip.

"We worked on it for several years," says Cluff. The scheme involved chartering a Russian nuclear ice breaker that was used to take tourists to the North Pole. Once the basics were worked out, they began recruiting members. "Steve was charging about $26,000 for a spot on the ship and he actually got about 40 people to put down the money."

Before the voyage, they chartered a plane to fly over the pole to locate the opening. "We were going to leave in August 2006. But in April of that year, Steve found out he had six inoperable brain tumours. Just before we were ready to fly, he died."

Another member of the expedition—Dr Brooks Agnew—was appointed as the new leader. After renaming the operation "The North Pole Inner Earth Expedition" and raising yet more funding, they planned for a summer 2014 departure. But a further unexpected disaster befell the team.

"Brooks Agnew resigned last September," says Cluff. "He said a major stockholder in his company had withdrawn all their money, saying it was because [Agnew] was involved in an expedition to find the Hollow Earth."

When another key member of the team died in an aeroplane crash, Cluff began to wonder if mysterious powers were manoeuvring against them. "There seems to be some force that's trying to stop this happening," he says. "I think it's the international bankers. They don't want the Inner Earth people messing around with their slaves, here on the outer world."[5]

Perhaps the strangest story of all is the one of "Dallas Thompson," who appeared on the late night *Coast to Coast* radio show with host Art Bell. Before the rise of the internet to propagate pseudoscience, paranormal ideas, and conspiracy thinking, Bell's show was conspiracy central during the 1990s and early 2000s. According to Storr,

Dallas Thompson was a former personal trainer who had spent his youth in Hawaii but now lived in Bakersfield, California. His life had changed forever following a terrible accident, five years earlier. He'd been driving along Highway 58 during heavy rain when his car had aquaplaned, spinning four times, only to plunge backwards down a 250 ft drop.

When Thompson was found, the roof of his blue Honda Accord had been crushed almost to the floor. The fireman who rescued him was amazed he hadn't been decapitated. As he'd been sitting, helpless, in the wreck, Thompson had had a vivid near-death experience. He claimed to have seen a "light so bright that it burnt my eyes" and made him "legally blind" and to have had bizarre knowledge about the world poured into him. When he regained consciousness, he was convinced that the Earth was hollow and had an opening at the North Pole. He'd come on Coast to Coast to discuss his mission to locate and explore it.

"There are cavern systems and caves that traverse the whole mantle," he told Bell, whose scepticism often took the form of slightly extended silences. Because of the special atmosphere in the hole, Thompson explained, living creatures were protected from pollutants and harmful rays. There were herds of mammoth and ancient tribes down there, the members of which lived to be around 1,700 years old. "How do you know all this?" asked Bell.

"I just do," said Thompson. "I remembered stuff that has been forgotten."

Later, Bell asked after his mental health.

"Are you manic?"

"I'm just excited," said Thompson.

". . . I can tell."

Perhaps most incredibly, Thompson revealed he'd secured funding to travel to the hole with a helicopter backpack called a SoloTrek, which he'd use to descend into it. He even had a date for the trip: May 24 2003.

Over the next few months, news of Thompson's expedition spread. He began to receive emails from media companies keen to report the story and many more from both critics and admirers. The sprawling book he'd written, which included his theories about Hollow Earth, began to sell.

In December 2002, two months after his radio appearance, he posted a message on his Yahoo Group page describing an inundation of "over 5,600 emails every few days." He said his book, *Cosmic Manuscript*, had become a bestseller but he was pulling it from sale. "I have requested the book be discontinued even though it's still at the top of the charts in Canada," he wrote.

And then, the most mysterious event of all took place. All of a sudden, Thompson disappeared.[6]

Storr tried to track down the mysterious Dallas Thompson.[7] He is apparently still living in Bakersfield, and his real name is Steven D. Thompson. He is now in his early fifties, but as his number had been disconnected, Storr's trail ran cold at that point. But there is no evidence of mysterious abductions and no corroboration of his incredible account.

How Do We Know about the Earth's Interior?

As in previous chapters, I have tried to explain to the reader *how* and *why* we know certain things about the earth rather than assert them as proven facts. The first problem with all these fantastic ideas about inner shells and caverns full of prehistoric beasts or the original humans is that as you descend into the *real* (rather than imaginary) earth, the temperatures grow hotter and hotter, and pressures get more and more extreme. The typical increase in temperature with depth (known as the geothermal gradient) is about 30°C for every kilometer you descend, so at 1.5 kilometers, the temperature is a scorching 45°C (113°F), at 3 kilometers, it is near the boiling point of water (almost 100°C), and at 30 kilometers, it is about 900°C, enough to melt most kinds of rocks.

This is why most mines must have good air conditioning as well as ventilation; rocks get hotter the deeper you get. Not only do the deep diamond mines of South Africa have powerful air conditioning but the conditions are so hot that the miners can work only in four-hour shifts before they are overheated and exhausted and must quickly ride an elevator to the surface to recover. I have personally experienced the geothermal gradient when, as a college geology student, I climbed down into the abandoned Mohawk Mine in the Mojave Desert. It got very warm once you descended into deeper and deeper shafts.

The pressure gradient is equally daunting. The pressure of the weight of all the rock above you increases by about 3 kilobars (3,000 bars of atmospheric pressure, or about 44,000 pounds per square inch) for every 10 kilometers (6.2 miles) you descend. Thus, even a small distance down into the crustal rocks of the earth, the pressures are enormous, and by the time you reach 30 kilometers (19 miles) down, the pressure is about 9 kilobars, or about 400,000 pounds per square inch. No human, or any device humans can build, could survive under such pressures and temperatures, despite the fantasies of the hollow-earth believers. Thus, the entire premise of the ludicrous movie *The Core* (starring Hilary Swank, fresh after winning an Oscar) is impossible; no device or craft could survive the temperatures and pressures even a few miles down, let alone drill down to the core itself. Hilary Swank and the entire cast would have been crushed into a pulp and vaporized after just a few miles of drilling. For these reasons, a poll of geoscientists voted *The Core* as the worst earth-science film ever made.[8]

Despite our best efforts and most sophisticated technology, the deepest humans have ever penetrated the crust is a borehole on the Kola Peninsula in Siberia. Even with the most advanced drilling technology and coolants to keep the drill bit from melting, the Russians were only able to penetrate down to 12 kilometers (7.5 miles) before the drill bit melted. In the 1960s, Project MOHOLE launched the first deep-sea drilling ship. They wanted to drill through the oceanic crust (about 10 kilometers thick), the thinnest crust on earth, to reach the mantle, but these efforts stopped due to the difficulties drilling through solid lava rocks of the oceanic crust, bureaucratic problems, and lack of funding. However, the current ship of the Integrated Ocean Drilling Project, the *Chikyu-Maru* (so huge it is nicknamed the "Godzilla Maru"), is designed to drill down through 10 kilometers of oceanic crust to the mantle in the near future.

For these reasons, the earth's interior cannot be inhabited by any creature, no matter what legends and science-fiction tales say. The earth's interior is completely inaccessible to direct sampling by scientists, except when volcanoes burp up a piece of mantle rock to the surface. Thus, everything we know about the earth's interior is determined by remote, indirect methods, primarily through geophysical observations. Such exploration has been going on for almost a century now, and we have learned a tremendous amount about the earth's interior, despite its inaccessibility.

1. **Gravity:** One of the best methods of determining the nature of the rocks beneath our feet is sensing the gravitational pull that they exert. For simple surveying of shallow crustal rocks, portable gravimeters (devices that measure the gravitational pull that they are experiencing) are used all over the world on a routine basis. For example, we can detect if there are large caverns beneath our feet, because instead of dense rock there is only air, so the gravimeter would give a reading that indicates much less gravitational attraction than average crustal rock. If, on the other hand, the material beneath you was a dense ore body or a mass

of denser mantle rock, the gravimeter would show that the density and attraction are stronger than for average crustal rock.

The same principle works for the gravitational attraction of the earth as a whole. Using Newton's law of gravitation, we can calculate the mass of the earth knowing the masses of other objects it interacts with, such as the sun, the moon, the satellites in orbit around us, and asteroids and comets affected by our gravity. The result from all these measurements is that the mass of the earth is about 5.97×10^{24} kilograms (that's 597 with twenty-two zeroes after it), or 1.32×10^{25} pounds. We also know the dimensions of the earth, so we can calculate the volume as 1.03×10^{12} cubic kilometers (1 million trillion cubic kilometers), or 2.5×10^{11} cubic miles. To get the average density of the earth, then, all you need to do is divide the mass by the volume, and you arrive at the value of 5.5 grams per cubic centimeter, or 5.5 times as dense as water. This number has been known since the 1774 Schiehallion experiment. By 1778, James Hutton had done calculations that showed, based on gravity and density, that the core of the earth must be a very dense metal.

Let's break that down further. The entire crust of the earth is a tiny part of the earth's total volume. To scale, the crust is about the thickness of wet toilet paper on a basketball. However, we know that the crust is much less dense than that global average, ranging from only 2.7 grams per cubic centimeter for granitic continental crustal rocks to 3.0 grams per cubic centimeter for basaltic oceanic crust. Mantle rocks, such as those that have come up to us from deep volcanoes, are only 3.3–5.5 grams per cubic centimeter in density. But the mantle is 84 percent of the total volume of the earth, so the fact that lowermost mantle density just barely reaches the global average of 5.5 grams per cubic centimeter means the core must be incredibly dense to make up the difference in order for the entire mass of the earth to be accounted for. The actual density values can be calculated to be about 10 grams per cubic centimeter in the outer core and 13 grams per cubic centimeter in the innermost core of the earth.

That's thirteen times as dense as water, so the pressure and density in the core is unimaginably huge! (Yet another reason why a movie like *The Core* was so ridiculous.) No matter what the hollow-earth believers wish to think, these numbers don't lie. We know from the density of rocks we can sample in mines and drill cores, and from these calculations, that there is *no* possibility of a hollow interior. On the contrary, the earth is made of dense silicate minerals in the mantle and even denser metallic iron and nickel in the core, with no place for any gaps or spaces.

2. Seismology: Thanks to news coverage during quakes, we are all familiar with how geophysicists use seismographs to detect earthquakes happening around the world. Within minutes, they can determine the size and location of almost any earthquake. But giant earthquakes produce enormous energy waves that travel through the entire earth and are detected by seismographs located all around the world. A really big quake sends out huge amounts of energy, so the earth is effectively ringing like a bell. Some quakes are large enough to even disturb the earth's rotation very slightly.

Since the 1920s and 1930s, seismologists have looked at all the earthquake records from seismic stations around the world and have obtained a three-dimensional picture of the structure and density of the earth based on these records. Up to 103° radial distance away from the earthquake, seismic waves travel in a curved path that eventually reaches the surface again and is picked up by a distant seismic station many thousands of kilometers from the original tremor (fig. 4.2). These waves have been bent, or refracted, as they travel through denser and denser layers of the mantle, just as light is refracted when it passes through the water in your fish tank and then back into the air or when the light is bent when it travels through the thick lenses of eyeglasses. Based on these wave paths, we can calculate that the base of the mantle, and thus the core-mantle boundary, is about 2,900 kilometers (1,800 miles) beneath your feet (fig. 4.2).

Beyond 103° radial distance from the earthquake, seismic waves do something even more interesting. The fastest are known as *P-waves* or *primary waves*, because they are the fastest waves of all and arrive first in a distant seismograph. They travel in a push-pull pulse motion like sound waves in air. From any given quake, P-waves travel in paths that show they have been refracted into a variety of complex paths as they traveled through the layers of the core, so they show up only at distances of 143°–180° from the original earthquake. It's as if the core were a giant seismic lens, focusing the waves in certain areas. Between 103° and 143° radial distance from the source, seismic stations do not pick up a given earthquake at all, because they are in the shadow of the dense

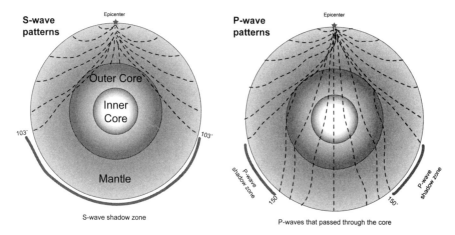

Figure 4.2. Seismic shadow zones. On the right, the waves from an earthquake on the North Pole would refract in broad arcs through the mantle up to about 103° radial distance from the epicenter. Beyond that point, seismographs would not pick up the P-waves from this earthquake until they were at least 143° radial distance away. The "shadow" of no arriving P-waves forms a ringlike zone around the opposite pole, caused by the denser core refracting the P-waves like a thick lens. S-waves, or shear waves, cannot pass through fluids. Like P-waves, they refract in large arcs through the mantle up to 103° radial distance from the epicenter. Beyond that distance, the entire opposite side of the earth doesn't detect S-waves from this earthquake, so it is an "S-wave shadow zone." In this case, the outer core is liquid, which cannot shear, so S-waves cannot travel through it. (*Redrawn from several sources.*)

rock of the core. The slower waves, or *S-waves*, not only arrive later up to 103° radial distance from the quake; they are also completely blocked by the shadow of the earth's core and do not show up at all beyond 103°. Since S-waves have an up-and-down shearing motion that cannot travel through fluids, we know that the outer core of the earth must also be a fluid, since it blocks S-waves from passing through it and thus casts a seismic shadow.

3. **Magnetism:** A third line of geophysical evidence is the earth's magnetic field. To generate such a magnetic field (see chap. 7), there must be a metal that is a good conductor, spinning around an axis through an existing magnetic field, which then generates a current and an even stronger magnetic field. This is the same principle that makes an electric motor work or the dynamo in a hydroelectric power plant generate electricity. From the evidence of the density of the earth's core, the behavior

of the earth's field over time, and meteorites, the only common metals available in the solar system that meet those properties are iron and nickel.

4. Meteorites: The fourth line of evidence is not geophysical; it's represented by visitors from outer space. Many types of rocks from space rain down on us, but for our purposes, there are two main groups that inform us about the earth's interior. Stony meteorites have a composition similar to rocks known to originate from the earth's mantle through deep volcanoes, so they are thought to be the original mantle material of small planets and large asteroids in space that were differentiated into a core and mantle. The rarest kinds are iron-nickel meteorites, which are composed of about 80–90 percent iron and 10–20 percent nickel. Probably from the cores of small planets and asteroids that broke up, they are the best proxy we have for the rocks of the earth's core, since they fit all the required proxies of density, magnetic behavior, and semifluid behavior indicated by all the lines of geophysical evidence.

Despite all this enormous amount of scientific evidence (mostly known since the 1930s), hollow-earthers continue to dwell in realms of thinking closely tied to the UFO believers and others who subscribe to the paranormal fantasy world. In this regard, they are different from the flat-earthers, the geocentrists, and the typical Young-Earth Creationists described in other chapters, who are largely motivated by religious beliefs and literal adherence to the word of the Scripture. However, like most of these other true believers, they are convinced that their belief is true and that it is being covered up by a giant global conspiracy among all the world's governments and scientists. That seems to be a common thread of nearly every weird idea about the earth.

5

Is the Earth Expanding?

Earth as an Inflating Balloon

Just one step less strange than flat-earthism and geocentrism is another in-ternet fad: the expanding earth model. Currently, it's got huge popularity due to a viral internet video by a cartoon artist, Neal Adams.[1] Using modern computer graphics, he put together a gee-whiz animation that appears to show all the continents fitting together in the past on a much-smaller globe. The video even plays the trite, overused opening chords of Richard Strauss's *Also Sprach Zarathustra* (famous from its use in Stanley Kubrick's film *2001: A Space Odyssey*) to create the appropriate sense of awe and wonder. If you read his writings further, Adams then takes the classic "fringe scientist" view of the world: all other scientists are wrong; they are in a great conspiracy to cover up the problems with their worldview and are under social pressure not to give his ideas a fair hearing. If you look closely at his arguments, how-ever, it is clear that he has no training or experience in geology or geophysics whatsoever and has no idea of the basic science of the earth he's trying to rearrange.

Back in the 1950s and 1960s, the expanding earth notion was still not considered outrageous. A prominent Australian geologist, S. Warren Carey, was the last legitimate scientist to be taken seriously on the idea of the ex-panding earth (his final book on the topic was published in 1975), since he pointed to the obvious fit between South America and Africa as well as the way that other continents fit together. But Carey's idea was soon overshad-owed by the immense amount of data that led to the modern theory of plate tectonics in the 1960s and 1970s, which not only explained Carey's fit of continents like South America and Africa pulling apart but also showed that other continents were colliding as tectonic plates converged and subducted beneath one another. When I was in graduate school in the 1970s, Carey's ideas were already an object lesson in the history of our science, a warning about how good observations can go badly wrong in the interpretation, especially when new data debunk an old notion.

Yet just as notions debunked over 500 years ago (flat-earthers and geocentrism) or 150 years ago (creationism) keep rising among those who know too little science, so too with expanding earth. Over fifty years since S. Warren Carey, the expanding earth notion has new life thanks to the internet and Neal Adams. And just like geocentrism and creationism, anyone with a few college courses in geology can easily debunk it, since Adams's cartoon completely ignores geology (which he has no experience in), just as creationist flood geologists ignore 99 percent of geology in order to explain the Grand Canyon in terms of Noah's flood.

How Do We Know?

As we did in other chapters, let's review the scientific evidence that falsifies the expanding earth idea. If you mistake Adams's slick cartoon for reality, it looks very convincing, although the video shows that there are lots of mismatches of the "fit" of the continents when slammed together that you don't notice, since the animation goes by very quickly and is very fluid.[2] But let's consider some actual geology rather than quibble over deceptive animations:

- The fit of South America and Africa is of course, real. Noticed by cartographers when the first good maps of the South Atlantic were published around 1500, the fit was used as evidence by the early advocates of continental drift, such as Antonio Snider-Pelligrini in 1885, Alfred Wegener and Alexander DuToit in the 1920s and 1930s, and S. Warren Carey (fig. 5.1). But the reason the fit is taken seriously is that there are geologic trends (such as scratches caused by glaciers scouring the bedrock) and identical rock types in the bedrock on each side of the Atlantic that match up when you push western Africa and eastern South America together to close the Atlantic and return it to its configuration about 250 million years ago. There are also matches in the bedrock between the rest of the Gondwana continents (India, Africa, Australia, and Antarctica) and bedrock similarities found in eastern North America and north Africa that formed before the North Atlantic opened up. But no such matches in bedrock geology exist between the other continents that Adams's animation squeezes together, especially those around the Pacific Rim. Geologic evidence shows that the Pacific has never been closed like the Atlantic; instead, it is a smaller remnant of the gigantic Panthalassa superocean that originally covered 70 percent of the globe about 250 million years ago, surrounding the supercontinent of Pangea.

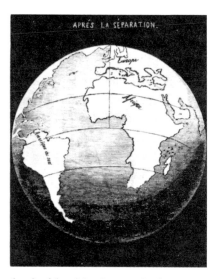

Figure 5.1. Antonio Snider-Pelligrini's famous sketch of the globe showing the fit between Africa and South America. (*Public domain.*)

- The way that the continents fit in Adams's animation is incorrect, since he matches the shape of the modern shorelines of the landmasses that are currently above water. The true edge of each continental plate is the edge of the continental shelf and the shelf-slope break, which is typically 1,500–2,000 meters below modern sea level. If you use this for the shape of the continent, the fit between South America and Africa improves (as geophysicist Sir Edward Bullard showed in the 1960s), but there is no fit for most of the other continents that Adams smashes together.

- The expanding earth model ignores a gigantic amount of paleomagnetic data collected from rocks of every age on every continent in the past fifty years. These data clearly show that the earth has not expanded more than 0.8 percent in the past 400 million years.[3] I've collected, analyzed, and published some of this data, and you can tell the paleolatitude of any given sample from the inclination angle of the specimens. If you look at all the samples of the same age over a range of latitudes (say, 250 million years ago in the mid-Permian when Pangaea formed), there is no possibility of a significantly smaller earth radius at that time, or any other. Also the paleomagnetic data give precise positions and orientations of each continent through the past six hundred million years, and these data do not support the fanciful motions suggested in Adams's cartoon.

- Plate tectonics has successfully explained a huge amount of data from biogeography and the ancient distributions of fossils as well, such as the distribution of Cambrian trilobites across the Atlantic, which would only work if the proto-Atlantic had closed in the late Paleozoic before reopening in the early Mesozoic. The strange motions of the continents in Adams's video do not explain these data in any sense.

- Plate tectonic models of the past motions of continents have successfully predicted where the climatically sensitive deposits of the world should be found in the past: glacial deposits on the poles, swamp deposits in the equatorial low-pressure belts, and desert deposits in the subtropical high-pressure belts. The cartoons by Adams fail to explain any of this.

- We can measure the diameter of the earth from hundreds of satellites with great precision, and these measurements show no evidence of the earth getting larger.[4] The satellite data are sensitive enough to see individual mountain ranges rising and local subsidence of basins, so we can detect uplift rates in the order of meters or less—and the same data clearly show the earth was not expanding within the decades that such data was collected.[5] At the rates of expansion suggested by the current expanding earth models, we should be able to detect such expansion even in a few decades.

- Using growth rings in corals and many other types of fossils that record the number of days in a year, we can easily calculate the gradual slowing of the earth's rotation due to the tidal friction from the moon's gravity. From this, we can estimate any changes in the earth's moment of inertia over the past five hundred million years, and there is no evidence that the earth has gotten any larger in that time.[6]

- Models of accretion or expansion on a scale required to significantly increase the radius of the earth do not match the known rates of accretion through geologic time. In addition, such expansion by accretion would release a lot of energy that would warm the earth's interior to a temperature much higher than it actually is.

- Models based on thermal expansion contradict the most basic principles of the rheology of the earth's interior and violate all sorts of constraints about the known mechanisms of melting and phase transitions within the mantle.[7]

- Even a beginning geology student can tell you that there are two kinds of fault systems: extensional faulting (found when the earth's crust pulls apart) and compressional faulting and folding (formed when the earth's crustal blocks collide to form mountain belts). Most of the world's great mountain belts (especially the Himalayas and the Alps but also, in earlier times, the Rockies and Appalachians) show clear evidence of having been formed by continental collisions as well as tremendous amounts of contractional folding, shortening, and faulting due to compression. If you look at any of the expanding earth models, the continents move apart but do not collide, and thus they fail to explain most of the world's mountains.

- Whenever you hear the expanding earth models explained, the proponents make a big fuss and argue violently that there is no subduction (the process whereby one tectonic plate slides beneath another and is remelted in the mantle). If they knew anything about earth science, they would realize that subduction is one of the best-documented processes in geology. Since the 1940s, seismic Wadati-Benioff zones have given us images of one plate plunging beneath another. The great Alaska Good Friday earthquake of 1964 first demonstrated one plate violently subducting beneath another, and the seismic evidence clearly showed that the plates were moving in the manner predicted by subduction.

  Since then, every earthquake on a subduction zone (including the big Sendai, Japan, quake of 2011) has shown similar behavior, and the seismic data clearly show the way the plates have moved. We can even use seismic imagery to see the plates sliding beneath one another.[8]

Without subduction, there would be no explanation not only for the seismic evidence of one plate plunging beneath another but also for the gravitational anomalies associated with subduction zones. Finally, there are many instances of ancient subduction zones that have been smashed into mountain belts and uplifted on land (as in the Coast Ranges of California). These ancient subduction zones have a characteristic suite of rocks, especially blueschist metamorphics, which could only be formed in the high-pressure but relatively cool regions of a subducting plate.

Expanding Earth Deflated

These are relatively simple problems with the expanding earth model based on basic geology that any advanced geology student could enumerate. There is a much larger problem that the expanding earth models fail to address: the source of energy to drive the expansion. Adams's video proposes simplistic ideas about physics and particle-particle interactions powering the idea of earth expansion (lampooned and debunked in this video).[9] This completely ignores the huge amount of evidence that shows that the earth's interior is not composed of hydrogen fusing into helium (as in the core of the sun) but has a core of both solid and liquid iron and nickel that is incapable of either expanding much or producing that much heat.[10]

Even more bizarre are the ideas of Stephen Hurrell of www.dinox.org, who not only advocates for the expanding earth but also claims that it might explain why the dinosaurs were so large; they lived in a world with much less gravity. As he writes,

> This is why the dinosaurs' large size is so interesting for the Expanding Earth theory. In 1987 I realized that the dinosaurs' large size could be explained by a reduced gravity and then soon realized that the most likely cause of a reduced gravity was a smaller diameter, less massive Earth. I'd never heard of the Expanding Earth theory before reasoning that the dinosaurs must live in a reduced gravity, but when I did learn about the geological evidence for an Expanding Earth it was soon evident that a Reduced Gravity Earth is exactly what an Increasing Mass Expanding Earth also predicted. Both theories use very different lines of reasoning and evidence but they both point to the same astonishing conclusions.[11]

Not surprisingly, many of the expanding earth websites are also creationist websites,[12] some of which use the "decreased gravity" of the earth to explain why dinosaurs could be so big! But this makes no sense at all. The gravitational attraction at the earth's surface depends on the mass of the earth below our feet, so unless they are advocating that the earth has also added mass somehow since the beginning of the Mesozoic, there can be no real difference in the gravity of the Mesozoic compared to today. If the mass is constant and all that increases is the diameter of the sphere, then it is density that changes. If these people are thinking that the smaller diameter means that dinosaurs were closer to the earth's center and thus the gravitational attraction is greater, they haven't done their math correctly. The difference in gravitational attraction is tiny between the Mesozoic earth they propose and today's earth.

Finally, the entire premise of their argument depends on the idea that all dinosaurs were so huge and massive that they would have had a hard time moving around in the modern earth's gravity field. But this is based on a grossly outdated idea of what the largest dinosaurs were like. We now know that they were much lighter than they appeared, because their bodies were full of hollow air pockets and tubes that made them relatively light. Not only that, but most dinosaurs were much smaller than the big sauropods, which their idea completely neglects. Finally, since the Mesozoic, there have been animals of dinosaurian size, such as the huge indricothere rhinos and giant mammoths, and they lived in gravity fields like we experience today. Thus, the entire idea is based on bad physics and math and a completely outdated idea of dinosaurs and other prehistoric animals.

In short, the ideas of the expanding earth advocates strikingly resemble those of the creationist flood geologists: they propose one simplistic model to explain a small part of the data and then ignore the other 99 percent of the data that don't fit. No one with even a rudimentary education in geology considers these ideas plausible, since they contradict so much of reality. Pushing the expanding earth as an unscientific mechanism to explain myths handed down from illiterate Bronze Age shepherds is certainly no way to enhance your credibility.

More importantly, the expanding earth model fails to explain the convergence of observations that has built and supported plate tectonics for the past seventy years. Like evolution or heliocentrism, plate tectonics is not just a single idea with only one line of evidence; it is a multifaceted theory that explains thousands of observations and makes successful predictions about new observations. After decades of such observations and the successful predictions that they produced, the burden of proof is on the expanding earth advocates to disprove every bit of evidence supporting plate tectonics. Likewise, after 150 years, the burden of proof is on creationists to explain away the entire edifice of observations and successful predictions that support evolution. And the tired old tactic of calling scientists "closed minded" and claiming they are "conspiring against" the idea is not going to convince anyone who knows the real data and the way that science (with all its internal criticism, peer review, and willingness to listen to crazy ideas that might be plausible) actually works.

A word of advice to Neal Adams: stick to cartooning. You're out of your depth in geology.

6

Did We Land on the Moon?

One Giant Leap for Mankind

July 20, 1969. Like many people born before about 1958 or so, I remember the day vividly, because it was a once-in-a-lifetime event that rivets your attention and crystallizes in your long-term memory where you were and what you were doing. I know exactly where I was when I heard the news of the assassination of President John F. Kennedy in 1963 (in my third-grade class, and another teacher came in crying and brought the news), where I was when Nixon resigned (in a college cafeteria at lunch during summer school; the entire place went wild cheering and celebrating when it came out over the public address system); when John Lennon was shot in 1980 (I was just five blocks away that night but didn't find out until the next morning at the newsstand, since I was a poor graduate student with no TV or radio then); when Reagan was shot in 1981 (I heard it from the lab people at the American Museum of Natural History in New York, where I was a student); where I was when 9/11 happened (just waking up in our old rented house in California, and my mother-in-law in Kansas calling us and telling us to turn on the TV); and several other major world events, such as the start of the first Iraq war.

I remember July 20, 1969, because I was fifteen years old and staying for the summer with my mother's cousin's family on their ranch in Oral, South Dakota, on the plains just east of the Black Hills and Hot Springs. I spent over a month there, getting exposed to ranching and farming life; learning to do farm chores, drive the tractor, ride horses, collect the eggs in the morning, and feed the chickens and hogs; and getting an education in things that most city kids never learn.

We were all gathered in their living room in front of their black-and-white TV, because the entire country had been following the progress of the Apollo 11 mission. We heard Walter Cronkite describe the events and the news coming from NASA Mission Control in Houston. Then they cut away

Figure 6.1. Iconic photograph of Buzz Aldrin on the moon, with astronaut Neil Armstrong and the lunar module reflected in the face shield. (*Courtesy NASA.*)

to the live video feed (a technological miracle for 1969) from the lunar module, and the entire world watched in real time as the first man set foot on the moon, the first time any human has ever stood on another body in space (fig. 6.1). We all heard Neil Armstrong speak those immortal first words on the moon in real time (with only a two-second delay caused by the transmission delay traveling the enormous distances from earth to the moon and back). The entire story of the mission has been brilliantly told by the recent movie *First Man*, and other parts of the space race have been dramatized in movies like *Apollo 13* and *The Right Stuff*.

Or at least that is how most people who are old enough remember that day and how history records the events. However, it seems that many major events (especially traumatic ones) also generate a large body of conspiracy believers with their own version of reality. Conspiracy thinking has over-shadowed the historic truths of the JFK assassination, the 9/11 attacks, the 7/7 terrorist attacks in London, the Holocaust, and many other well-doc-umented events. Sure enough, there are a small but significant number of people who believe that the moon landing and the entire space program were an expensive hoax and that the moon missions were all faked and filmed on a Hollywood soundstage.

When most people hear this (especially if they are old enough to re-member witnessing the events live), they are stunned. The entire space pro-gram was an expensive hoax? Hundreds of thousands of NASA employees and contractors, all the astronauts, and the entire government managed to stage this with complete secrecy, and no one has ever leaked once? For many people (including myself), the first reaction is revulsion that people's imag-inations are this wild and their belief system so warped that they would fall for a silly conspiracy theory like this. Strictly speaking, the moon landing conspiracy is not a weird idea about the earth but about the moon and the way we got there. Nevertheless, it is worth including in this book because it is much like the other weird ideas, conspiracy theories, and bizarre views about our planet and solar system that we have encountered.

This particular notion was apparently started in 1976 with a self-pub-lished book, *We Never Went to the Moon: America's Thirty Billion Dollar Swindle*, by Bill Kaysing, a former Navy officer with a limited science or space technology background. Between 1956 and 1963, Kaysing had worked as a writer for Rocketdyne in Chatsworth, California, where the F-1 engines used on the Saturn V rockets were first built. (There is a claim that back in December 1968 there were already conspiracy theories in circulation about the Apollo 8 mission as it became the first to circle the moon.[1]) The idea gained momentum in the 1970s as the country became disillusioned, after the lies surrounding Watergate and the Vietnam War caused people to lose faith in their government.

In 1978, Peter Hyams directed the film *Capricorn One*, which depicts a faked landing on Mars with a spacecraft that looked just like *Apollo*, creating the public notion that an entire space program could be a hoax. As Stanley Kubrick's landmark 1969 film *2001: A Space Odyssey* (based on the book by

Arthur C. Clarke) demonstrated, it was possible to create a believable moon walk in a studio soundstage. In 1980, the Flat Earth Society accused NASA of faking the moon landings, claiming they were filmed by Disney based on a script by Arthur C. Clarke and directed by Stanley Kubrick!

In 2002, *Dark Side of the Moon*, a French mockumentary about Kubrick filming the moon-landing footage, was released as a spoof of the moon-landing hoaxers and immediately taken to be a genuine documentary by the conspiracy crowd! Folklorist Linda Dégh wrote that "the mass media catapult these half-truths into a kind of twilight zone where people can make their guesses sound as truths. Mass media have a terrible impact on people who lack guidance."[2] After that, the conspiracy theory took off, so now there are dozens of books, videos, and even films by many other people with no background in space science or technology and an alarming number of websites on the topic in the cesspool of lies that is the internet.

Even more alarming is how much this bizarre conspiracy theory has gained traction and been accepted by a significant number of Americans. In 1994, the *Washington Post* found that 9 percent of its respondents thought it was possible that the moon landing was a hoax, and another 5 percent were unsure.[3] A 1999 Gallup poll found that 6 percent doubted that the moon landing happened and 5 percent were unsure,[4] consistent with a 1995 Time/CNN poll.[5] The situation was not helped when the Fox network ran the special *Conspiracy Theory: Did We Land on the Moon?* in February 2001, reaching fifteen million viewers. Subsequent polls showed that about 20 percent of Americans thought the moon landings had been faked.[6] Astronomer Phil Plait, who often writes about pseudoscience, reviewed the program as follows: "In one word, it would be 'grotesquely distorting reality, an execrable steaming pile of offal that doesn't come within a glancing blow of the truth.' Was that more than one word? Well, it's hard to find a single word that truly captures the feel of that program."[7] It's bad enough that American television runs a constant stream of pseudoscience, from UFO shows to ghost hunters to Bigfoot shows, just to gain an audience, but the media also give credibility to the conspiracy theorists with shows like this.

Another survey showed that about 25 percent of Russians (who are far more conspiracy-minded after years of government lies and propaganda) also believed the moon landings were faked.[8] And a 2009 poll of British citizens showed that about 25 percent believe the conspiracy theory about the moon landings.[9] Clearly, the media seems to feed the conspiracy lunacy rather than doing the responsible thing and running specials that demonstrate scientific evidence for the moon landing. About the only positive media influence is the brilliant show *Mythbusters*, which ran an entire hour-long episode debunking the moon-landing conspiracy myth. The complete episode is available online through vimeo.com (just do a browser search for "Mythbusters moon hoax" and the Vimeo video will come up).[10] The episode is worth watching in its entirety, not only because it is entertaining but also because they do an effective job of debunking many of the common claims of conspiracy theorists.

There is no positive evidence to support the different claims of how the moon landings were supposedly faked. Over four hundred thousand people worked on the program, NASA employees and others, most of whom did not sign any kind of secrecy pledge, and no one has ever come forward and claimed that they weren't actually working on rockets or space technology. My dad worked for Lockheed from 1938 until he retired in 1977, helping build every aircraft and spacecraft they produced during those years. He frequently told my family when he was working on NASA projects. No secrecy documents ever restricted him. (He also had top-security clearance to work on the Lockheed spy planes in Area 51 and never told us anything about it until he was retired and the secrecy restrictions were no longer binding. He found the crazy ideas about Area 51 hilarious, because he was there, and there were never any alien bodies or UFOs, just top-secret spy planes and stealth aircraft.) No person who was actually working for the space program has come forward with tales of how it was faked.

There are many different versions of how the moon landings were allegedly faked, and none of them agree on many details except that the landings occurred in a movie studio, not on the moon. Instead of offering a robust, testable scientific explanation of their own, the moon conspiracy theorists take the same tactic as creationists: try to nitpick details of the documented NASA records on the assumption that if any one little piece of the puzzle doesn't fit, the entire "hoax" collapses like a house of cards. But just as in the case of the creationists, the argument doesn't work that way. NASA has provided a mountain of evidence, along with the experiences of four hundred thousand workers and dozens of astronauts, documenting what they did. The burden of proof is on the conspiracy nuts to show that *all* that evidence is false or faked and that *all* those people are lying; one little problematic piece of evidence does not prove a conspiracy existed.

Myths Busted

Since the entire case for the moon-landing conspiracy rests on the nitpicky details of specific technical and scientific issues, we will consider them individually. Let's first look at the silly claims of the conspiracy theorists and evaluate why they are garbage.

1. **Van Allen radiation belts:** Conspiracy theorists claim that the astronauts could not have survived their trip through the Van Allen radiation belts. In fact, the metal alloy (mostly aluminum) skin of the *Apollo* spacecraft was sufficient to block most of the ionizing radiation, and the trajectory of their launch and trip to the moon took them through one of the thinnest parts of the Van Allen belts. This myth is busted by none other than Dr. James Van Allen himself, who wrote in response to that Fox "documentary" mentioning this issue: "The recent Fox TV show, which I saw, is an ingenious and entertaining assemblage of nonsense. The claim that radiation exposure during the Apollo missions would have been fatal to the astronauts is only one example of such nonsense."[11]

2. **Oddities in photographs:** Conspiracy theorists pore over NASA photographs, claiming to find oddities that they think suggest that the images were shot in a studio and not on the moon's surface. However, each of these supposed anomalies has been examined by independent photography experts with no connection to NASA, and they all have been explained by more mundane causes. For example, the crosshairs of the camera appear to be behind objects on the moon, but this is clearly an artifact of the emulsion of the film bleeding over the tiny fine lines on the photograph because the photos were overexposed. If you look at the original high-quality NASA images, rather than poor low-resolution copies, this effect vanishes.

Others point out the lack of stars visible in the sky; they claim that NASA chose not to put stars in the studio backdrops so that astronomers would not be able to detect whether they matched with the real sky or not and thus discover that the landing was a fake. But the conspiracy nuts underestimate how bright the light is on the moon's surface at daytime; it is at least as bright as a sunny day on earth, and we can't see most stars in the daytime here either. The cameras are set to get proper exposures for the bright objects in the foreground, so they would not be sensitive to dim points of light in the background. Even on earth, to get good images of stars, you need a very dark night away from light pollution of cities as well as a fairly long time exposure—none of which *Apollo*'s cameras were set up to photograph. In fact, there *are* images taken by *Apollo*'s far ultraviolet camera/spectrograph, which did record star images in the UV spectrum, and those images match the pattern of stars seen on earth and by space telescopes.

One set of myths concerns a mysterious *C* shape that seems to be drawn on a moon rock. However, this only appears on the copies; the original image doesn't show this *C*, which suggests that a thread or hair fell on the enlarger or negative when the copy was made. (Those of us who did photography in the age of film remember how often dust and hairs could get on the enlarger lens or the negative and spoil the print.) Other conspiracy theorists claim that the lighting and shadows are inconsistent with the lighting on the moon. What they fail to take into account is that the lighting on the moon is fairly complicated. It comes not just from the direct sunlight on one side of an object or astronaut; there is also reflected light from the moon's bright surface, and wide-angle lenses introduce distortions. In some cases, the light is scattered by lunar dust that was kicked up by the astronauts, and this dust takes a long time to settle, since the moon's gravity is only one-sixth the strength of the earth's gravity. (This particular myth was busted in the *Mythbusters* episode mentioned earlier.)

3. The moon's surface was so hot in the daytime that the camera film would have melted: Without an atmosphere to transfer heat by conduction and convection, the moon transfers heat from its surface by direct radiation, and both the astronauts and their cameras were well shielded against heat. In addition, nearly all the moonwalks were timed to occur just after the lunar sunrise or just before the lunar sunset, when the surface is at its coolest during the daytime. Those time intervals are long, since a moon day lasts about fifteen earth days.

4. The flags don't look right: One of the most famous and most ridiculous claims is about the flags planted on the moon during several of the Apollo missions. Claims have been made that the flag doesn't hang properly for zero gravity or that images show it fluttering when there is no air on the moon to move it (fig. 6.2). First of all, the flag is standing out straight from the vertical flagpole, because the upper end of the

Figure 6.2. Buzz Aldrin planting the American flag on the moon. (Courtesy NASA.)

pole is bent into an *L* shape, so the flag is hanging from the horizontal crossbar. There are no actual video sequences showing the flag fluttering, other than when it was still slowly unfolding from the tightly folded position after it was hung on the L-shaped frame. The "ripples" on the flag are not caused by any kind of wind but by the fact that the flag was tightly folded in storage and made of a stiff fabric, so it remained wrinkled and rippled even after it was unpacked. The motion also continues much longer than expected because there is no atmosphere to slow the flapping down, whereas on earth, it would have stopped fluttering much sooner. (This was another false idea effectively debunked in the *Mythbusters* episode.)

5. The footprints on the moon seem too crisp if there is no soil moisture to hold them together: Actually, moon dust has never been weathered (since there is no water or oxygen on the moon) and has sharp edges, so it sticks together by static electricity and its own angular grain shape and holds its shape in a vacuum (fig. 6.3). The astronauts

Figure 6.3. Photo of his own footprint taken by Buzz Aldrin on the moon. (*Courtesy NASA.*)

themselves compared the moon dust to "talcum powder or wet sand" in its behavior.[12] (Yet another canard effectively debunked in the *Mythbusters* episode.)

6. The moon landings were filmed on a soundstage or out in the remote desert: This is easily debunked by a variety of observations. First, the motion of the dust from the lunar rovers could only happen in a vacuum and in a gravity field one-sixth as strong as that on earth. Second, on the Apollo 15 mission, astronaut David Scott did a version of Galileo's famous cannonball experiment. As the video shows,[13] a hammer and a feather dropped at the same time reach the ground at the same time (just do a browser search for "Apollo hammer feather" and you will find it). This could only happen in a vacuum—air resistance would have slowed down the feather. If the landings were filmed in a desert, the heat waves of the air rising from the desert surface would be visible in the videos, but there are no such heat waves. (This entire claim was also debunked in the *Mythbusters* episode.)

7. The lunar modules did not make blast craters as they descended to land on the moon: Actually, during the final stage of descent, the lunar module was moving quite slowly and the descent engine only had to support the craft's weight, which was decreased by the moon's weak gravity, so the engine thrust is only ten kilopascals (1.5 PSI).[14] The engine exhaust gases spread broadly out of the nozzle; as gases expand much more quickly in a vacuum than in an atmosphere, there was never a concentrated narrow blast of gas as they descended. And if you watch the NASA footage closely, the engines did scatter a lot of the fine lunar dust when they landed, and the astronauts commented on this fact in their radio communications with NASA. Finally, the bedrock of the moon is very close to the surface, since there is no weathering to make a thick soil, like earth's, so the lander would hit bedrock quickly and not sink very deep. In addition, the coating of dust is very shallow, so the weak gases of the descending lunar module could never blast a crater out of solid moon rock.

8. The lunar modules made no mark on the moon's surface, yet the much-lighter astronauts left distinct footprints: The lunar module may have weighed 17 tons (15.3 metric tons) on earth, but it weighed only 2.6 tons (1.2 metric tons) in the moon's weak gravity with almost all its fuel burned up. The lunar module also had very broad footpads designed to prevent them sinking very far, somewhat like snowshoes, while the astronaut's boots were much smaller and narrower, so they put many more pounds per square inch on the surface when they walked. In fact, there are some photographs where you can see the footpads of the lunar module pressed down into the lunar dust, especially when the vehicle moved sideways as it touched down.

This is just a small sampling of the silly ideas that have been proposed about the moon landings and the simple answers to all these supposed oddities. There is no point in belaboring this point any further. Let's move on to the positive evidence for why the moon landings are indeed real.

How Do We Know?

How do we know that the moon landings really happened as described? What is the positive evidence?

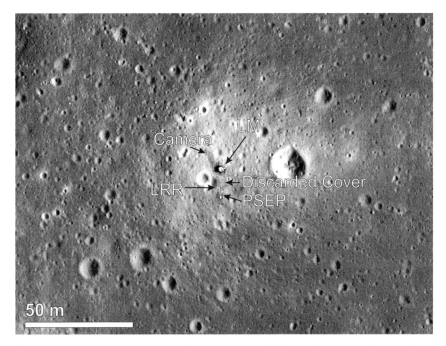

Figure 6.4. Image shot by the Lunar Reconnaissance Orbiter in March 2012 from only 24 km (15 miles) above the surface, showing the Apollo 11 landing site and the material left after the astronauts departed; LM = Lunar Module; LRR = Laser Ranging Retroflector (and the discarded cover for that instrument); PSEP = Passive Seismic Experiment Package, designed to detect moonquakes. (*Courtesy NASA.*)

1. The surface of the moon is continually mapped in 3-D by satellites, and the topography they recover matches the NASA Apollo photographs down to the finest detail. Many other spacecraft have flown over the various Apollo landing sites since 1972, and they have photographed the materials left behind by the Apollo missions, sometimes many years after the astronauts left. The Lunar Reconnaissance Orbiter has circled the moon since 2009. Not only has it photographed the leftover equipment, but it also has the resolution to see the footprints (fig. 6.4)!

2. The moon dust flying off the wheels of the lunar buggies travels in a way that could only happen in a vacuum with gravity about one-sixth that of earth. If it were faked on a soundstage, the dust would swirl in a vortex as it does behind a car driving on a dirt road, since the earth has

a dense atmosphere and gravity is six times stronger here than it is on the moon. Instead, the NASA images show the dust rising and falling in simple parabolic arcs, exactly as predicted by simple calculations that anyone who knows a bit of physics could perform if they wanted to analyze the Apollo film footage. In fact, this calculation has been performed many times. The effect could only happen if there were no atmospheric interference and the gravity was much weaker than that of earth.

3. Several of the Apollo missions placed small retroreflectors on the equipment left behind so that they could be tracked at a later date. Since then, a number of experiments have successfully bounced laser beams off those reflectors and back to earth. The Laser Ranging Retroflector experiment successfully sent out laser beams and picked up the reflectors on the Apollo 11 equipment on August 1, 1969. The reflectors on the Apollo 14 craft were detected by lasers from McDonald Observatory on the same day it took off from the moon. The reflectors on the Apollo 15 lander were picked up by McDonald Observatory just a few days after it left the moon. All these experiments were to test whether there were any changes in the distance between earth and the moon since the reflectors were left, but they also confirm that NASA astronauts were there. The Russians also landed unmanned craft on the moon and have successfully picked up reflections of lasers from them. (*Mythbusters* does an effective experiment and successfully detects a laser bounced off the moon.)

4. We have thousands of samples (about 2,200 total, plus a lot of moon dust) of lunar rocks that are unique and cannot be mistaken for any earthbound rock. I've had the opportunity to look at several of them myself. Any geologists worth their salt could tell a moon rock at a glance and recognize how it differs from all known terrestrial rocks. Moon rocks show no evidence of weathering in a wet environment like earth, have unique geochemical signatures, and many of them date from 3.9 to 4.6 billion years old, older than any rock known on earth. In some cases, there are individual distinctive rocks that we can match to the footage and images from NASA.

For example, a distinctively shaped moon rock nicknamed "Big Muley" is the largest rock ever brought back from the moon (fig. 6.5). A football-sized, twenty-six-pound stone that was recovered by Apollo 16 astronaut Charles Duke is on public display. You can see it clearly in

Figure 6.5. The lunar rock sample nicknamed "Big Muley," collected by astronaut Charles Drake during the Apollo 16 mission and now on display at NASA. (*Courtesy NASA.*)

the NASA photos from where it was recovered from the rim of Plum Crater in the Descartes Highlands. There is also NASA footage of Duke picking it up, carrying it, and commenting on how large and heavy it is.[15] All this has been verified by third-party scientists, since many scientists from countries other than the United States have done research on moon rocks, and they certainly aren't part of any US government conspiracy.

Some conspiracy theorists try to dismiss this evidence by claiming that these are meteorites blasted off the moon that reached earth. If they had ever seen meteorites, they would have noticed a distinctive fusion crust that formed from heating up as they plunged through earth's atmosphere, and any competent geologist could tell they were meteorites. Not only that, but the first lunar meteorite wasn't found until 1979, and its lunar origin wasn't established until 1982. Only 30 kilograms of lunar meteorites have ever been found after decades of searching, yet the Apollo missions brought back 380 kilograms of moon rocks in just seven missions and only a few man-hours over just two years. To collect that much lunar meteorite material from the earth's surface, many man-years would be required.

5. Even more convincing, the Soviets tracked the Apollo missions with their own technology, and never once did they suggest that there was anything amiss. Remember, we were in a space race with them at the time, and if they had any doubts that our missions were real, if they thought they might be hoaxes, they would have tried to humiliate us with a big propaganda blast (as they did when they shot down our U-2 spy plane in 1960). Instead, they said almost nothing, because we had beaten them to the moon fair and square. There are even images of the front page of *Pravda* posted online that show the Soviet government acknowledging the US success.

There are some who claim that both the United States and Soviet Union conspired together to cover up what happened, but anyone who knows the nature of the rivalry during the Cold War would realize at once that this is ridiculous. They were our enemies; they had no reason to cooperate with us and every reason to try to embarrass us if they could. Instead, they acknowledged that they had been beaten to the moon. The Soviets had their own successful unmanned lunar missions, and as I write this book, the Chinese have successfully landed an unmanned probe on the far side of the moon. These events could not be faked when all these world governments are competing and have no reason to engage in a global conspiracy.

6. Over five hundred thousand people (including my dad) worked on various parts of the Apollo program over a period of about ten years. Over one hundred thousand people on the ground (plus millions more on live television) watched the Apollo 11 Saturn V rocket blast off at Cape Canaveral, Florida, a few days before it reached the moon, and similar numbers watched every other launch. What were all these people doing and seeing if we were engaged in a giant hoax? And how is it that *not one of them has come forward* and claimed that they were hoaxes?

Professor James Longuski of Purdue Aeronautics and Astronautics Engineering pointed out that it would have been easier and cheaper to actually land on the moon than to keep a conspiracy of this size going for over fifty years now.[16] As we discussed in chapter 1, this is the fatal flaw of any giant conspiracy theory. People just don't keep secrets very well, and the more people who try to cover something up, the more likely that one of them will eventually leak the truth. Besides, none of these people signed secrecy agreements keeping them from talking, since most of the work of NASA is public record and is still available to anyone who wants to do research. Sure, the Nixon White House and the Central Intelligence Agency had some secrets, but they all ended up being leaked.

NASA was *not* a secret organization, and no employees were restricted in their ability to talk about what they were doing, except that they were not supposed to publicize technical details that might give the Soviet space program some of our technological advantages. In 2005, on an episode of their Showtime series *Bullshit* (season 3, episode 3), magicians Penn and Teller covered the moon-landing conspiracies. As Penn Jillette pointed out, the sheer number of people involved in NASA and the Apollo program greatly exceeded the Watergate conspirators—yet one was leaked right away, and the other has never seen a break in the "conspiracy."

This is not an exhaustive debunking of all the claims made by the conspiracy crowd, since that would take a book by itself and the details are all spelled out on www.clavius.org. But it is enough to see how ridiculous the moon-landing conspiracy sounds to any fair-minded skeptical observer. There is no need to belabor the point further here. Let's just say, in the words of *Mythbusters* hosts Adam Savage and Jamie Hyneman, "Myth *busted!*"

Magnetic Myths

Flip-Flop Fakes

All throughout the long buildup to the latest failed prediction of a global apocalypse in 2012, you would hear people claiming that the earth-shattering catastrophe on December 21 would include "pole shifts" or "changes in the earth's magnetic field" and all sorts of other sciencey phrases proclaimed by people with absolutely no idea what they were talking about. The idea of magnetism is one of the most popular memes in the lexicon of pseudoscientists and New Agers, since magnets seem to operate "mysteriously" and exert a force at a distance. From the days of the quack and charlatan Franz Mesmer claiming he had "magnetism" over people to the trite phrase *animal magnetism*, the concept of magnetism has always been mysterious and misunderstood. Hence the big market for sticking magnets on various parts of your body to "cure" you. All they do is waste money and possibly demagnetize the magnetic strip on your credit cards. The ideas that somehow the earth's magnetic field will shift abruptly, the earth's core will stop rotating (as in the laughably bad Hilary Swank movie *The Core*), and, even more wildly, the earth's rotational pole will change are all common out there in the paranormal world.

Among these crazy ideas is the notion that somehow the magnetic poles will shift and destroy all electrical devices, thus destroying civilization.[1] One website claims that pole shifts will cause earthquakes and hurricanes and that NASA is covering up what's happening.[2] Another site cherry-picks items from actual science posts and then completely misinterprets what they mean.[3] This quote is typical:

> Magnetic pole shifts are in the news a lot recently because our Magnetic North Pole is racing towards Russia while the Earth's magnetic field strength is falling fast. The European Space Agency said this could be the beginning of a magnetic pole shift. Evidence suggests the rotational axis

will also shift to stay aligned with the magnetic field. These catastrophic pole shifts come in a periodic cycle of recurring and predictable cataclysms involving huge earthquakes and tsunamis, changes in latitude and altitude, mass extinctions, and the destruction of civilizations—reducing them to myth and legend. Most people don't want to know this is coming. They don't want to know that America could replace Atlantis in legends of lost civilizations. Your government censors the evidence to make sure you'll keep working and funding their preparations until it happens. But for those who are willing to look, the evidence exists.[4]

This is just a small sampling of the pseudoscientific garbage all over the internet posted before December 21, 2012. And the failure of the latest apocalyptic prediction in 2012 hasn't stopped these ridiculous claims; that quote is from April 2018. Most of us know enough about apocalyptic predictions to guess that they are not worth taking seriously, but very few people have bothered to debunk this stuff. Unfortunately, lots of people did take the ridiculous apocalyptic predictions seriously, often with tragic results.

Among my other specialties, my professional training is in paleomagnetism, and I've conducted over thirty-five years of published research in the field, so I'm pretty familiar with what we do and do not know about the earth's magnetic field and how it behaves.

How Do We Know?

The earth's magnetic field has at least two components: the *dipolar field* (fig. 7.1), which makes up about 90 percent of the magnetism we normally feel, and a *nondipole field*, which is normally hard to detect but makes up about 10 percent of the earth's field. The dipole field is not exactly lined up with the rotational axis of the earth (i.e., there is a small angle between magnetic north and true north), but over geologic spans of time, this magnetic north wanders around the rotational pole in a movement known as secular variation.

Studies have shown that over the long term, the position of magnetic north averages out to be identical to the rotational pole. The field is generated by complex fluid dynamos operating within the outer core of the earth (made of iron and nickel), which operate a bit like a spinning dynamo made of copper wire (a good conductor) that generates a magnetic field and electrical current when it spins within a magnetic field. Exactly how this works

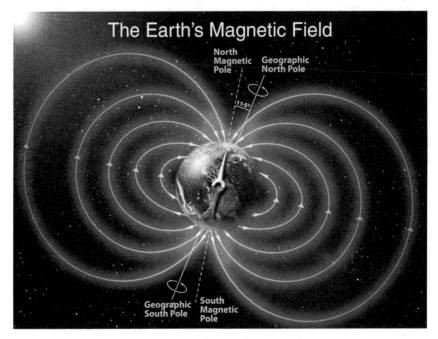

Figure 7.1. Diagram of earth's magnetic field. (*Redrawn out of copyright.*)

is a matter of the complexities of geophysical fluid dynamics, so scientists are still working on modeling what kinds of dynamos are found in the outer core. But whatever their configuration, the model has to fit the constraints that the direction can be reversed (so a compass pointing north now would point south eight hundred thousand years ago) and must also explain the odd behavior of the field when the dipole component weakens and the non-dipole component becomes visible.

Now let's consider some of the common false claims that plague the internet.

1. *The earth's field is about to reverse!* The earth's field does reverse direction, but normally the process takes between four thousand to five thousand years to complete. It does not happen in days or weeks, as some claim. We know this from detailed studies of thick stacks of lava flows that erupted over the span of a reversal, as well as high-resolution deep-sea cores that accumulated millimeters of sediment for every few

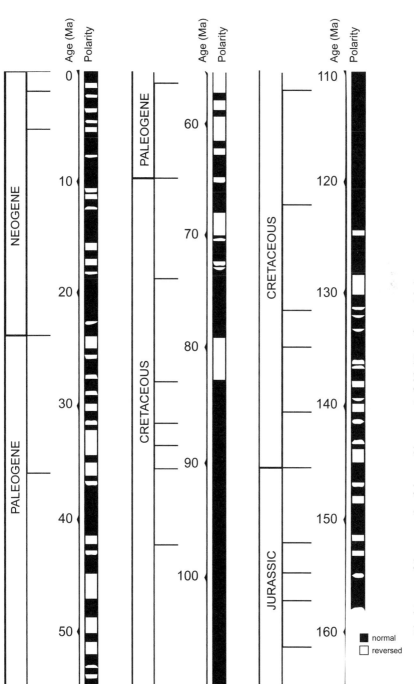

Figure 7.2. The history of the reversals of the earth's magnetic field. (*Out of copyright.*)

years and span the interval. So if the field were beginning to reverse, we would not know for at least a few thousand years. And we cannot predict when this reversal will occur, since they have been occurring on an irregular basis for all of geologic history—at least three hundred times in the past one hundred million years. Reversals typically occur roughly two hundred thousand to three hundred thousand years apart, although the last reversal (the Brunhes-Matuyama boundary) was over eight hundred thousand years ago. Some, however, are much shorter (less than fifty thousand years), while in other cases, the earth's field remained stable for thirty million years. This irregular pattern of field reversals (fig. 7.2) is completely unpredictable, but it also gives a nice nonperiodic, nonrepeating signal, analogous to a bar code, that allows magnetic stratigraphers to correlate their local magnetic sequences with the global pattern.

2. *When the field reverses and vanishes, we'll all be bombarded by cosmic radiation!* Actually, when the field slowly reverses over thousands of years, only the dipolar component of the field weakens. The nondipole component of the field is always present, and there's no evidence that the earth has ever been unshielded from cosmic radiation or completely lacked a magnetic field. Nor is there any evidence that a slightly weaker magnetic field over the thousands of years when the dipole field is reversing will have any effect on life, on our electrical grid, or on anything else.

Calculations show that during reversal, the field is only slightly weaker than we feel normally—about the difference between the field we would feel at the equator and the field we would feel at the magnetic north pole. In other words, it is undetectable except by sensitive instruments. In fact, Jim Hays of Lamont-Doherty Earth Observatory, a micropaleontologist, my coauthor on several papers, and a former professor of mine, conducted the crucial experiment on the issue over forty-nine years ago.[5] He was the first to see the evidence of field reversal in deep-sea cores from the Antarctic, and he wondered if there was any effect on life. Hays's first effort in 1971 demonstrated no association between field reversals and extinctions, and it has since been corroborated over and over again by a wide variety of statistical techniques.[6]

And why should there be any extinctions? If the difference in the field felt by organisms is so slight and the effect on the cosmic ray influx is so tiny, there's no reason to expect otherwise. At the very worst, a weaker magnetic field with a relatively strong nondipole component might disorient animals (from bees to whales to birds) that navigate by the dipolar field direction, but there's no way to test that hypothesis in the fossil record and no evidence that it's happening to organisms right now.

3. What about recent studies that showed much more rapid field changes? Some of these were conducted by my former colleague Scott Bogue at Occidental College. Bogue was looking at a set of lava flows that cooled as the field was reversing and weakening, and he focused on just the field recorded when the dipole field was nearly gone and the nondipole field was revealed. The nondipole field does indeed move rapidly and in weird ways, but there's no evidence that anything has been affected by such weird field directions during the short period of time that the nondipole field is dominant. And there's no evidence that the much-stronger dipolar field will ever change that quickly.

4. What about the evidence that the magnetic pole is rapidly changing direction? This is a long-studied and well-known phenomenon called secular variation, as I have already mentioned. It's not news, nor is it some scandalously dangerous discovery being hidden by NASA. Secular variation is a constant feature of the earth's magnetic field, but over time, the direction of the magnetic pole averages out to be approximately the same as the rotational pole. We can study secular variation over thousands of years as recorded in deep-sea cores, lake sediment cores, and many other records. There is no scary change that threatens us, just a lot of noisy wobbling of the magnetic north that averages out to nothing in the long run.

So the next time you hear some "prophet" worrying about the earth's magnetic field, you can assume that it's misinformed and false. There are plenty of real dangers to worry about, like global climate change, so we don't need to scare people by hyping false ideas.

Earth-Shaking Myths

*If you want to understand geology, study earthquakes. If you
want to understand the economy, study the Depression.*
—Former Federal Reserve chair Ben Bernanke, "The Great Depression Lesson 1"

*Southern Californians freak out when it rains, yet when
there's an earthquake they're like "pass the salt."*

—Gregor Collins, *The Accidental Caregiver*

Quaking in Their Boots

Of all the natural disasters that humans experience, earthquakes are scarier to more people than any other. Many people who have never experienced one are often deathly afraid of them, even though quakes are extremely unlikely to kill anyone in the United States thanks to our building codes and construction. (The same is not true of many underdeveloped countries in Asia, where the loss of life can be extreme.) Charles Darwin noted this in his visit to Chile on the *Beagle* voyage: "The earthquake, however, must be to every one a most impressive event: the earth, considered from our earliest childhood as the type of solidity, has oscillated like a thin crust beneath our feet; and in seeing the laboured works of man in a moment overthrown, we feel the insignificance of his boasted power."[1]

I've run into all sorts of people terrified of quakes that they have never actually felt, yet they don't even flinch at much deadlier events like hurricanes and tornadoes. There are all sorts of legends associated with quakes, from "earthquake weather," to the idea that California will fall into the sea (no, it's sliding north to Alaska at a few centimeters per year on average), to the myth that fault lines look like huge, deep chasms floored with lava, as in the first Christopher Reeve *Superman* movie (no, they just form straight valleys on the ground; chasms are due to landslides occurring far from the fault). I've lived through every major Southern California quake since I was born, including the 1971 Sylmar quake, the 1987 Whittier quake, and the 1994 Northridge quake. When I was in New York in August 2011, I even got to experience a rare eastern quake (the Virginia quake that shook the tenth floor of the Frick Wing of the American Museum while I was visiting), and then, ironically, I had to leave later that week and cancel my talk to NYC Skeptics on my new book on natural disasters because Hurricane Irene was on the way.

Quakes are probably particularly scary to people for two reasons: they are unpredictable (unlike weather events, which give some warning), and they shake our confidence in terra firma, which we have all grown up to assume cannot move. Psychologists have shown that human beings are notoriously poor at judging relative risks and assessing which threats are really serious and which ones are exaggerated. For deeply held psychological reasons, people are far more afraid of dying from a snakebite or in an earthquake, even though both of these events are staggeringly improbable for most people in the United States.

Only five to ten people die of a snakebite each year, and earthquakes have killed an average of only six people per year in the past century in the United States. Yet because of irrational psychological reasons, we are unjustifiably afraid of them. Because snakes trigger a primordial fear response in our brain, we are terrified of them. When we were small, vulnerable hominids running across the African savanna, snakes were a real threat to us, because many African snakes, like mambas and cobras, are venomous. But now that snakes are so heavily slaughtered in this country (despite the fact that most American snakes are not venomous), we are much more a threat to them than they are to us.

The best way of assessing real threats is to look at cold hard statistics, as an actuary or insurance adjuster does. An article by Borden and Cutter looked at deaths in the United States from all natural hazards from 1970 to 2004.[2] Despite the fact there were several big California earthquakes (1971, 1987, 1989, 1994) and large hurricanes during that time window, you would never guess what the number one killer among natural disasters was. It was not even a topic that we think of as catastrophe, since it happens so often and so slowly. The top killers among natural hazards in this country are heat waves, storms, and winter!

Yes, earthquakes, hurricanes, and other extreme weather events are terrifying disasters, but the biggest killers are slow and subtle: heat waves. Likewise, we take severe storms and the bitter cold of winter for granted since they happen so often, but they kill a lot more people than more dramatic events like tornadoes, hurricanes, and earthquakes. It turns out that hurricanes, earthquakes, and landslides are near the bottom of the list of relative risk, with less than 2 percent of total deaths. Even though hurricanes and tornadoes are potentially very dangerous, we usually have some warning of when they are coming, and most people take shelter or evacuate when warned. Volcanic events did not even make the list, since the small Mount St. Helens eruption was the only deadly volcanic event in this country for better than a century.

Let's put that in an even broader perspective. Many people are terrified of earthquakes and tornadoes and hurricanes, but these events are not something to lose sleep over, except when there are clear warnings that a hurricane or tornado is coming. We should be more careful and worried about heat waves and severe winter storms, but we're so accustomed to these each year that we don't realize how deadly they are. Worrying about natural disasters looks absurd in the face of where the real risks come from: your cheeseburger and french fries, your car, cigarettes, and all sorts of things you encounter every day.

Borden and Cutter point out that for the 20,000 people killed by natural disasters in the United States during the study period of 1970–2004, there were 652,000 deaths from heart disease alone (more than thirty times the natural disaster total)! There were 600,000 deaths from cancer (also thirty times the total from natural disasters). Of cancer deaths, almost a third were from lung and other cancers due to smoking. Colorectal cancer, pancreatic cancer, prostate cancer, and breast cancer were the other biggest killers. There were 143,000 deaths from stroke, 130,000 from chronic lower respiratory diseases (bronchitis, pneumonia), and even 117,000 deaths from accidents (mostly car accidents).

If we really took the issue of risk seriously and evaluated it objectively, we would do well to improve our diet and exercise, get frequent health checkups, stop smoking, and modify our driving habits. We may fear death in an earthquake or hurricane, but lunch, cigarettes, and driving are much deadlier to us!

The Myth of "Earthquake Weather"

As St. Patrick's Day was beginning to dawn on March 17, 2014, in Los Angeles, most people in the central and western side of the city were abruptly awakened just after 6:30 a.m. by a 4.4-magnitude quake on the north side of the Santa Monica Mountains, near Woodland Hills, where I used to teach geology at Pierce College. It awoke my family during the two to five seconds of shaking, but I slept through it. I've been through every major quake here since I was born in the region, so a little 4.4 event doesn't even rattle me.

We immediately turned on the news and were told basic information about the quake; the magnitude was downgraded from 4.7 to 4.4 when better data came in, and the location was moved from Westwood to Sherman Oaks, closer to the actual fault line on the north side of the Santa Monica Mountains. Most of the reporting was competent, although in the early stages, it's largely nonscientific stuff like "Did you feel it?" and "What did it feel like?" rather than anything accurate or scientific that would tell us something important about the quake. Sure enough, sooner or later, it was bound to happen: one of the man-on-the-street interviewees spouted the geologists' least favorite myth: "Oh, it was warm yesterday, so there must have been earthquake weather." Fortunately, the news anchor was smart enough to dismiss this urban myth and move on to another interview, but if you surfed the internet, it was full of claims that earthquake weather must have caused this quake. (Geologists' other pet peeve is when people—especially news anchors and reporters—use the term *tidal wave* for tsunamis, which have nothing to do with tides. Fortunately, the tragedy of the December 26, 2004, Sumatran quake and tsunami seem to have reduced the incidences of these displays of ignorance.)

The myth of earthquake weather goes back all the way to Aristotle, who thought that earthquakes were generated in underground caves as the air was trapped during hot, sultry days and supposedly shook the earth when it swirled around in the cave. Like most of the fanciful notions that Aristotle dreamed up about nature, this myth persisted for over two thousand years; the monks faithfully copied the ancient texts, and the Church came to regard Aristotle as the final authority on nature, even when he was clearly and demonstrably wrong. Not until Newtonian physics and Darwinian biology came along were most of Aristotle's false ideas about science gradually debunked and replaced by scientific explanations from actual observation and experiment, not intuitive fantasies. It was not until the 1906 San Francisco quake that it was finally proved that earthquakes are caused by movement on fault lines, not air moving in and out of caves, or the wrath of God, or any such other mythic idea.

The stories of earthquake weather are a classic example of false correlation. People are hypersensitive to everything that happens during a traumatic earthquake, and especially if the weather is hot and sultry and uncomfortable just before the quake, they make the false connection that the weather caused the quakes. According to W. J. Humphreys back in 1918, earthquake weather is a psychological manifestation. Humphreys argued that "the general state of irritation and sensitiveness developed in us during the hot, calm, perhaps sultry weather given this name, inclines us to sharper observation of earthquake disturbances and accentuates the impression they make on our senses, so that we retain more vivid memories of such quakes while possibly over-looking entirely the occurrences on other more soothing days."[3]

The earthquake weather myth also demonstrates confirmation bias. We remember when one or two quakes happened during "earthquake weather" but fail to notice the weather during most of the quakes we feel. But the data on earthquakes have been analyzed hundreds of times, and there is absolutely no correlation between any weather phenomenon and the occurrence of quakes. Earthquakes happen around the clock every few minutes somewhere, and in active areas like California, we get many small quakes in a day and large ones every few years or so somewhere in the state. But we never feel most of these quakes, so we never notice them. Yet one minor quake happens after a hot day and people are immediately blathering on and on about earthquake weather.

In fact, if you think back to the last few major quakes in the Los Angeles area, none happened during hot weather spells. The January 17, 1994, Northridge quake happened at 4:30 a.m. on a cool winter day, similar to the February 9, 1971, Sylmar quake, which hit at about 6:00 a.m., also in winter. The October 1, 1987, Whittier quake also hit early in the morning, at 7:42 a.m. In fact, I can't think of any recent major quake in my part of the world that struck during a hot summer or autumn day. If one wanted to make a false correlation with weather, our biggest local quakes seem to occur in the early morning during the winter, the exact opposite of the conventional scenario. But of course, that is silly too. Earthquakes happen around the clock, year in and year out, during every season and every kind of weather condition, so there is no pattern whatsoever. If you buy into the idea that earthquakes happen in the early morning, you run into another problem. As this source explained it,

> While there have been some memorable quakes that fit the dawn time-frame (e.g. the 1994 Northridge quake, a 45-second 6.7 shaker at 4:31 a.m. on 17 January 1994 and the (estimated) 7.9 that took apart San Francisco at 5:12 a.m. on 18 April 1906), there have been many others that haven't. The 10 March 1933 6.4 magnitude Long Beach quake hit at 5:55 p.m., and the 18 May 1940 Imperial Valley 6.9 quake struck at 8:37 p.m. And the 17 October 1989 Loma Prieta 7.1 shaker happened at 5:04 p.m., wiping out parts of the Nimitz freeway just as commuters were driving home from work.
>
> Even more revealing is the fact that in different cultures, there are completely different kinds of "earthquake weather," so that just about ANY weather pattern is thought to cause earthquakes somewhere on earth![4]

If you think about it, there is no physical basis to believe in earthquake weather for one simple reason: the changes of atmospheric temperature over the course of hours to days don't penetrate more than a few meters underground, while the faults that cause earthquakes are many kilometers underground and cannot be affected by the changes of temperature under any imaginable circumstances. As humans who live in the bottom layer of the atmosphere, we are overly impressed by and sensitive to weather, and we think of its power extending everywhere. We fail to realize that underground is an entirely different world, much bigger and more thermally stable than above ground's rapid changes in atmospheric gases. To better grasp the concept, just think of all the burrowing animals in the desert that can escape killer heat with burrows just a foot or two beneath the surface, and you can better realize why earthquake weather makes absolutely no sense.

However, the research continues. A couple of claims have been made that huge hurricanes and tropical cyclones can increase the frequency of aftershocks on a quake, but these have not yet been corroborated on a more rigorous basis—and there is no known mechanism for even something as strong as a hurricane to penetrate the ground deep enough to trigger fault movement.[5] The scientific community hasn't stopped looking at the possibilities, but one or two preliminary studies demonstrating a possible correlation aren't enough to overcome the huge body of evidence showing that there is no connection between the vast majority of quakes and any weather condition.

For now, however, the myth of earthquake weather belongs with other urban myths, such as the idea that there are alligators in the New York sewer system. Don't believe everything you read and hear—especially when it seems to be a popular legend!

9

Quacks and Quakes

Earthquake "Prophets"

The great Japanese earthquake and tsunami of March 11, 2011, not only generated huge coverage in all the media but also brought all the crazies out of the woodwork (as any major earthquake or natural disaster does). Each time something like this happens, we are inundated by a wave of publicity for all the cranks who claim they predicted the quake. Among those who got their fifteen minutes of fame during the postquake media blitz in 2011 was a well-known crank, Jim Berkland, who on March 17 got a long interview on Fox News (but on no other network).[1] First, the reporter put up a map of the Ring of Fire of volcanoes and earthquakes around the Pacific Rim; pointed at Chile, then New Zealand, then Japan; and implied that this circle of quakes might end in California. Apparently, he never consulted a geologist, who would have pointed out that each of those regions is an entirely different type of plate boundary and they have no tectonic plates in common.

Then Fox gave Berkland a full five minutes to spout his ideas, with the same credulous reporter tossing him softball questions and no rebuttal from any other geologist or seismologist. Berkland rambled on about animal behavior and fish die-offs in California (which have been explained by unusual water conditions), never mentioning that such die-offs are common and not statistically associated with earthquakes (nor is there any plausible mechanism that might link them). He blathered on about how animals sense unusual magnetic fields before the quake, an argument that has been thoroughly debunked, and he mentioned several other quakes that he claimed to have "predicted," with no fact-checking or examination of his overall record of "prediction." He also demonstrated the classic persecution complex of all cranks and fringe scientists, dismissing real scientists and their "black boxes" when he uses methods with no rigor or peer review. He then used his airtime to boldly predict that a great California quake would happen on

March 19, the perigee of the supermoon, with a prediction window running to March 26. Well, those dates came and went over eight years ago, and we never saw Fox News interviewing him again to explain what went wrong.

If Fox had bothered to do minimal research about him, they would never have wasted the airtime and panicked people unnecessarily. First of all, Berkland is not a seismologist but a geologist with only a bachelor's degree and some graduate training who served in several different government positions before retiring in 1994. Berkland is touted as having predicted the Loma Prieta earthquake, something that he uses as his main publicity hook and that was promoted on the Fox broadcast. However, there are questions about this prediction. That region had already been targeted a year earlier as a seismic gap, one of the most likely areas for the next big quake. According to the US Geological Survey, "The segment of the San Andreas fault that broke in the 1989 M 7.1 Loma Prieta or 'World Series' earthquake had been identified by the USGS as one of the more likely segments of the San Andreas to rupture. Magnitude 5+ earthquakes 2 and 15 months before the damaging earthquake were treated as possible foreshocks, and the USGS issued 5-day Public Advisories through the California Office of Emergency Services."[2] So it was no great shakes to follow this prediction and pick a date. Berkland just got lucky and happened to mention his date to a reporter for a local paper in Gilroy, California, so it was actually placed in the public record.

Once his complete record of quake prediction is examined more closely, its success rate falls apart. It's a classic case of cherry-picking favorable data as well as confirmation bias, used by fortune-tellers and faith healers and swindlers of every kind for centuries: people remember the hits and forget the misses. Berkland got one lucky hit, and most people now never bother to check his overall track record. Seismologist Roger Hunter did a careful statistical study published in the *Skeptical Inquirer* and found that Berkland's predictions were no better than chance.[3]

Berkland claims that when the tidal forces of the alignment of the moon and sun are at a maximum, they can exert a pull on the earth's crust and trigger earthquakes. This idea goes back to at least 1897, and it has some plausibility, since tidal attraction does exert some force on the earth's crust. However, when geophysicists at the UCLA Institute for Geophysics and Planetary Physics conducted a rigorous study, they found no statistical relationship between the two.[4] Several other studies have also tested the connections between tides and earthquakes, and they have found no statistically significant correlations.[5] The only possible correlation might occur when tidal forces pull on shallow thrust faults, but the faults in Berkland's March 19 prediction for California are all deep, vertical strike-slip faults; none has the type of motion that fits the shallow thrust-fault model. The fault that caused the Sendai quake is a deep subduction zone, not a shallow thrust. The only other possible place where tides might affect quakes is in the rift valleys of mid-ocean ridges, miles under the middle of the ocean and far from the areas that Berkland has focused on in his predictions.[6]

In addition to questionable methods and lack of consistent success in prediction based on tides, Berkland uses animal behavior as a guide to predicting earthquakes. His "highly rigorous" method is to survey the newspaper for an unusual number of lost dog and cat reports in the classified ads section. The idea that animals can predict earthquakes has been carefully analyzed and has failed the test again and again.[7] Animals may be more sensitive than humans to P-waves, the fastest seismic waves that arrive several seconds before the destructive S-waves in regions far from the epicenter, but this gives warnings of only a few seconds to a few tens of seconds in any place that is likely to experience strong shaking. If animals are sensitive to other disturbances in the earth's crust that happen more than a few seconds before the quake itself, it has never been reliably corroborated. In fact, the cases of "odd animal behavior" are also tainted by confirmation bias. If you watch animals long enough, you'll see many episodes of "odd behavior"—and then you promptly forget about it. But if there is an episode of "odd behavior" just before a quake, your mind immediately makes the false association between the two (just as in the case of "earthquake weather" debunked in chap. 8)—and you completely forget all the times you observed "odd behavior" and no earthquake occurred. In addition, this method runs into the same problem that most short-term earthquake-prediction methods have encountered: no two earthquakes are alike. Some have precursors, and others don't. Thus, if animals did act strangely before a particular quake occurred (just as some geophysical precursors have been observed on some quakes), there is no evidence that they reliably predict most quakes (just as many quakes don't have precursors).

This leads to the bigger issue: the best seismologists in the world have been working hard on short-term earthquake prediction for decades, but most would concede that we are not much closer than we were fifty years ago. We are very successful at giving long-term warnings of months to years in advance for regions that are overdue for a big quake (seismic gaps), and these predictions have worked reliably. But short-term prediction has always foundered on the maddening problem that no two faults behave in the same way.

Back in the 1970s, dilatancy theory about ground deformation found a series of precursors, and that led to the successful prediction of the February 4, 1975, Haicheng quake in China. But just seventeen months later, there were no precursors for the July 28, 1976, Tangshan earthquake in China, and about half a million people died. Since this failure, seismologists have become much more cautious about short-term earthquake prediction. Most will candidly admit that there will probably never be a reliable method of short-term prediction. This leaves room for quacks like Jim Berkland to step in, brag about his questionable Loma Prieta "prediction," and get free media attention. Then he can rely on the fact that reporters these days do no research into his background, nor do they confront him after each failed prediction to ask him what went wrong. As Charles Richter himself said, "Only fools, liars, and charlatans predict earthquakes."[8] You can be the judge of which category best fits Berkland.

Following False Prophets

We have been discussing when society is conned by false earthquake prophets. The flip side is the problem of when a society takes those prophets seriously and then punishes people who warned against them. This occurred on April 6, 2009, when a Richter magnitude 5.8 (moment magnitude = 6.3) earthquake struck the province of Abruzzo in central Italy. It killed 309 people, injured 1,173 more, and 65,000 people were made homeless. The quake damaged almost eleven thousand buildings in the medieval city of L'Aquila and caused about $16 billion worth of damage over the region. This was the deadliest earthquake to hit Italy since the 1980 Irpina quake, a Richter magnitude 6.9 event in southern Italy, which killed 2,914 people, injured over 10,000, and left 300,000 homeless. The L'Aquila event was preceded by hundreds of foreshocks, which caused much of the population to flee the city and seek shelter before the main quake. There were also hundreds of aftershocks, some of which were over 5.3 in Richter magnitude.

Naturally, people were upset and wanted someone to blame. In the case of most natural disasters, people usually regard such events as "acts of God" and try to get on with their lives as best they can. No human cause is responsible for great earthquakes, tsunamis, volcanic eruptions, tornadoes, hurricanes, or floods. But in the bizarre world of the Italian legal system, six seismologists and a public official were charged with manslaughter and convicted for not predicting the quake![9] My colleagues in the earth science community were incredulous and staggered at this news. Seismologists and geologists have been saying for decades (at least since the late 1970s) that short-term earthquake prediction (within minutes to hours of the event) is impossible, and anyone who claims otherwise is lying. How could anyone then go to court and sue seismologists for following proper scientific procedures, let alone convict them?

What's going on here? There's more to the story, of course. Apparently, an Italian lab technician (*not* a qualified seismologist) named Giampaolo Giuliani made a prediction a month before the quake based on elevated levels of radon gas.[10] However, seismologists have known for a long time that radon levels,[11] like any other magic bullet precursor, are unreliable because no two quakes are alike and no two quakes give the same precursors. Nevertheless, his prediction caused a furor before the quake actually happened.

The director of the civil defense, Guido Bertolaso, forced Giuliani to remove his findings from the internet (old versions are still online). Giuliani was also reported to the police for "causing fear" with his predictions about a quake near Sulmona, far from where the quake actually struck. Enzo Boschi, the head of the Italian National Geophysics Institute, declared, "Every time there is an earthquake there are people who claim to have predicted it. As far as I know nobody predicted this earthquake with precision. It is not possible to predict earthquakes."[12] Most of the geological and geophysical organizations around the world made similar statements in support of the proper scientific procedures adopted by the Italian geophysical community. They condemned Giuliani for scaring people using a method that has not been shown to be reliable.

Sadly, most of the press coverage I have read (including many cited above) took the sensationalist approach and cast Giuliani as the little David fighting against the Goliath of Big Science. Apparently, none of the reporters bothered to do any real background research or to consult with other legitimate seismologists who would confirm that there is no reliable way to predict earthquakes in the short term and that Giuliani was misleading people when he said so. Giuliani's prediction was sheer luck, and if he had failed, no one would have mentioned it again. Even though he believes in his method, he ignores the huge body of evidence that shows radon gas is no more reliable than any other predictor. In this regard, he is much like other quack "scientists" like Jim Berkland, who get free news coverage predicting earthquakes and then the press never bothers to challenge their credibility or ask the quack, "What happened?" when his prediction later proves false. People want to believe that solitary geniuses are better than the hundreds of scientists who have established a large body of evidence and research and that his treatment was due to his success, not to his crying wolf.

So it came as a shock when, on October 22, 2012, the Italian courts convicted six scientists of manslaughter for failing to predict the earthquake. Seismologists around the world were stunned that scientists were demonized for doing their jobs properly, even though the defense had offered numerous witnesses from the international seismological community who testified that short-term quake predictions are impossible. I've read through all the accounts I can find, but it seems that the testimony of the international seismologists was completely ignored. Instead, the trial focused on the efforts of the scientific officials to prevent widespread panic by asserting that Giuliani's prediction had no scientific basis (which is true) and that there was no strong evidence of a major earthquake coming soon (also true). The courts seemed to be punishing scientists for their efforts to prevent panic and for realistically stating that the probability of Giuliani's prediction being correct was very low. Unfortunately, when the quake did happen, people tried to find someone to blame, and scientists were convenient scapegoats.

As University of Southern California seismologist Tom Jordan wrote,

> The Italian scientists were trapped by a simple yes-or-no question: "Will we be hit by a damaging earthquake?" This was not surprising given Giuliani's alarms, but it was not one they could answer conclusively. From what they knew a week before the earthquake, a big shock was not very likely: the probability of a false alarm (if an alarm were raised) exceeded the probability of a failure-to-predict (if an alarm were not cast) by a factor of more than 100. Even so, seismic activity had increased the probability of a large earthquake by a significant factor, perhaps as much as 100-fold, above the long-term average. Distracted by Giuliani's predictions, the authorities did not emphasize this increase in hazard, nor did they focus on advising the people of L'Aquila about preparatory measures warranted by the seismic

crisis. Instead, they made reassuring statements that were widely interpreted to be categorical.[13]

Fortunately, the scientists' conviction was overturned on appeal in 2014,[14] after over five thousand seismologists wrote an open letter to the courts and to Italian president Giorgio Napolitano, pointing out that earthquake prediction is a fantasy and that the accused did the proper thing by trying to prevent panic after the unqualified lab technician had issued a spurious (but accidentally correct) prediction. Although the people of the L'Aquila region were not happy with the appeal verdict and protested, the scientific community in Italy and around the world rejoiced.

Stefano Gresta, president of Italy's National Institute of Geophysics and Volcanology, said, "The credibility of Italy's entire scientific community has been restored."[15] The seven scientists and officials were freed, but they needlessly suffered almost five years of pain and anguish as they were falsely accused, convicted, and imprisoned for doing the scientifically responsible thing, while the Italian courts and a lot of people bought in to the lies spread by a quack technician. As a number of people pointed out, the quake was not that big, but the ancient flimsy construction of the buildings in L'Aquila made its effects much worse. An official in Italy's Civil Protection Agency noted that "in California, an earthquake like this one would not have killed a single person."[16] Since the 1933 Long Beach quake and the Field Act, all of California's buildings have had to meet a strict set of construction standards to make them resistant to destruction in earthquakes.

This raises another question: What does this imply for scientists who are working in a field that might have predictive power? In a litigious society like Italy or the United States, this is a serious question. If a reputable seismologist *does* make a prediction and fails, he or she is liable, because people will panic and make foolish decisions and then blame the seismologist for their losses. Now the Italian courts are saying (despite worldwide scientific consensus) that seismologists are liable if they *don't* predict quakes. They're damned if they do and damned if they don't. In some societies where seismologists work hard at prediction and preparation, such as China and Japan, there is no precedent for suing scientists for doing their jobs properly, and the society and court system does not encourage people to file frivolous suits. But in litigious societies, the system is counterproductive and stifles research that we would like to see developed.

What seismologists would want to work on earthquake prediction if they could be sued? I know of many earth scientists with brilliant ideas about not only earthquake prediction but even ways to defuse earthquakes or to slow down global warming, or many other incredible but risky brainstorms, but they dare not propose the idea seriously or begin to implement it for fear of being sued. This state of affairs sure isn't good for science or for society. But until the more litigious countries find some way to address the problem, potential advances that scientists could make to improve our lives are unnecessarily held back.

(10)

Was There a Great Flood?

In the Beginning . . .

Nearly every ancient culture that lived near a river had legends about a great flood. In many cases, the flood washed away all their records, destroyed most of their property, and killed so many people that it became a legend that also washed away any memory of previous history. There are flood stories from the Greeks, the Indus River peoples, the Chinese, the Egyptians, and many others. Very few of these stories have anything in common besides the flood theme, so there is no reason to think they are describing a single worldwide event. Instead, this commonality reflects how completely a huge flood wipes out any memory of the preflood past and forces people to reconstruct their history in legends.

One of the oldest flood myths that still survives is *The Epic of Gilgamesh*, which dates to about 2750 BCE. The Sumerians had a hero called Ziusudra (called Atrahasis by the Akkadians and Utnapishtim by the Babylonians) who was warned by the earth goddess Ea to build a boat. Tired of the noise and troubles of humanity, the god Ellil planned to wipe them out with a flood. When the floodwaters receded, the boat was grounded on the mountain of Nisir. After Utnapishtim's boat was stuck for seven days, he released a dove, which found no resting place and returned. He then released a swallow that also returned, but the raven he released the next day did not return. Utnapishtim then sacrificed to Ea on the top of Mount Nisir.

Since the Hebrews spent much time in captivity in Babylon and were influenced by the neighboring Mesopotamian cultures (especially if Abraham came from the Sumeria, as the Bible says), it is not surprising that they adapted this well-established Mesopotamian legend for their own mythology. The story is nearly identical to that of Noah's flood, not only in its plot and structure but even in the details of its phrasing. Only the characters' and gods' names and a few details have been changed to suit the differences between the monotheistic Hebrew culture and the polytheistic cultures of the Sumerians, Akkadians, and Babylonians.

Two centuries of biblical scholarship have shown that the Bible is a composite of many different sources, written at different times by different authors, all with different motivations. They are distinguished by the use of certain key phrases and words and the style of the Hebrew in which they were written. I learned to read Hebrew when I was in high school, so I soon discovered what a mishmash of stories the Bible really is. For example, the account of Noah's flood in Genesis 6 and 7 is actually two different accounts interwoven almost verse by verse; as a result, they are often not consistent with each other, and in some places the two accounts directly contradict each other.

One source is called the J source, after Jahveh or Yahweh (written *YHWH*, since it was forbidden to speak God's name aloud). Jahveh was a common Hebrew name for God, later misspelled and mispronounced "Jehovah" by later authors who filled in vowels between the consonants that were never in the original. (The Hebrew system of vowel points was not yet invented.) The authors of the J document were priests of the southern kingdom of Judah, who wrote sometime between 848 BCE and the Assyrian destruction of Israel in 722 BCE. The other main source is the P source, or Priestly code, apparently written by priests of Aaron about the time of Babylonian captivity in 587 BCE. Between 622 to 587 BCE, the Hebrews began to intercalate the different sources, so they became all mixed together. This did not bother them, since they didn't take the Bible as a literal document but a spiritual one. Only modern fundamentalists, who don't read the Bible in the original Hebrew, make the mistake of taking it literally.

This is apparent when you read the Noah's flood story with one of the many modern scholarly Bibles that identifies the source of each verse. Genesis 6–7 gives the story of Noah twice, once from the J source and once from the P source, with verses from the two sources intermingled, so they sometimes contradict each other. Genesis 6:5–8 is from the J source, but Genesis 6:9–22 is from the P source. Then Genesis 7:1–5 is from the J source, but Genesis 7:6–24 is alternately from the J and P source every other line or so. This leads to many contradictions, such as Genesis 7:2 (from the J source) saying that Noah took seven pairs of each clean beast in the ark, but Genesis 7:8–15 (from the P source) says that he took only one pair of each beast in the ark. In Genesis 7:7, Noah and his family finally enter the ark, and then in Genesis 7:13, they enter it all over again (the first verse from the J source, the second from the P source). According to Genesis 6:4, there were Nephilim (giants) on the earth before the flood; then Genesis 7:21 says that all creatures other than Noah's family and those on the ark were annihilated, but Numbers 13:33 says there were Nephilim after the flood.

The story of Noah's flood was intended as a spiritual message to the Hebrew people to remind them that they were chosen descendants of Noah and that God had made a covenant with them. Unfortunately, by the Middle Ages, it had become a literal doctrine in most of Christian Europe, since they did not know about the sources of the Bible and their inconsistencies as we do now. Yet even back in 408 CE, the great Christian scholar St. Augustine warned Christians against taking the Bible too literally. In his words,

> It not infrequently happens that something about the earth, about the sky, about other elements of this world, about the motion and rotation or even the magnitude and distances of the stars, about definite eclipses of the sun and moon, about the passage of years and seasons, about the nature of animals, of fruits, of stones, and of other such things, may be known with the greatest certainty by reasoning or by experience, even by one who is not a Christian. It is too disgraceful and ruinous, though, and greatly to be avoided, that he [the non-Christian] should hear a Christian speaking so idiotically on these matters, and as if in accord with Christian writings, that he might say that he could scarcely keep from laughing when he saw how totally in error they are. In view of this and in keeping it in mind constantly while dealing with the book of Genesis, I have, insofar as I was able, explained in detail and set forth for consideration the meanings of obscure passages, taking care not to affirm rashly some one meaning to the prejudice of another and perhaps better explanation.[1]

The message, however, was lost on centuries of European scholars, right up until the early and middle 1800s. At that time, both the detailed analysis of the Bible in the original Hebrew and the discoveries of scientists discredited literal interpretation of the Scriptures. The detailed analysis of the Hebrew documents was performed by a series of mostly German Hebrew scholars and came to be known as Higher Criticism. It forms the basis for all modern Bible scholarship.

Science and Noah's Flood

The scientific case against Noah's flood, however, came from natural philosophers (as they were then known) starting in the late 1700s. At first, they interpreted the rock record in terms of the flood myth. According to scholars like Giovanni Arduino in 1759, the "Primary" or "Primitive" rocks were hard granitic rocks and metamorphic rocks like schist and gneiss, supposedly formed during the earth's creation. Above them were the "Secondary" rocks, which were hard layered sandstones and limestones, full of fossils deposited on the flanks of mountains; in some interpretations, they were formed in Noah's flood. Above them were "Tertiary" rocks, poorly consolidated sands and gravels, often on the foothills of the mountains. In the view of some scholars, these were deposited as the flood receded, while others considered them flood deposits. They also recognized the "Diluvium" or "Drift" (later called Quaternary) deposits, loose sands and gravels along with giant out-of-place boulders (which we now know are glacial deposits of the Ice Ages).

Prominent German scholars such as Abraham Gottlob Werner of the Freiburg Mining Academy lectured in 1775 and afterward that the layered rocks of the earth had all been deposited in water in a global flood. Werner was not necessarily referring to Noah's flood, although most scholars did tie this interpretation to the biblical story. Nonetheless, Werner was adamant that all layered rocks (including what we now know are lava flows) were originally laid down in water. For this reason, they came to be known as the Neptunists, for the Roman name of the god of the sea. You might ask how anyone could think that lava flows were formed in water rather than by flowing molten volcanic rock! Remember that in the late 1700s, none of these scholars had ever traveled far from home. More to the point, none had ever seen an active volcano erupt and release lava flows. By contrast, nearly everyone today has seen many videos of eruptions of Kilauea volcano in Hawaii or other volcanoes. It was not until the early 1800s that the chemistry of the minerals in a lava rock was understood, and it was clear they were once molten lava not material formed in water.

The answer to Werner's dogmatic approach to the earth came from the "Father of Modern Geology," a Scottish gentleman-farmer named James Hutton. He lived in the peak period of the Scottish Enlightenment during the 1760s through 1790s, when learning and scholarship blossomed free of the restrictions of any church or dogma. Gentlemen often met in drinking clubs around Edinburgh and Glasgow, and some of them were legendary geniuses, including not only Hutton but also Adam Smith, whose book *The Wealth of Nations* described modern capitalism; David Hume, the famous philosopher and historian who wrote some of the earliest philosophical treatises on skepticism and religion; James Watt, the inventor of the first practical steam engine, which helped launch the Industrial Revolution; the famous chemist Joseph Black; and many others. Hutton, Smith, and Black were particularly close friends, forming a drinking society known as the Oyster Club, which held wide-ranging and free discussions about any and every topic, from science to politics to religion to philosophy.

Hutton was fascinated by how the earth worked. Trained in both law and medicine (and making a tidy profit from his skills as a chemist), Hutton was a wealthy gentleman who didn't have to work for a living. He had the time and money to be able to afford to think about the processes at work in the soils and erosion of his farms across Scotland. Unfettered by the strictures of religion or Wernerian dogmas of Neptunism, he came to look at the earth in a truly original way. He found numerous cases of lava flows in Scotland that had clearly forced their way up through the layered sedimentary rocks as a hot intrusion of liquid rock, melting the rock around them as they did so. Thus, the Wernerian dogma that lava flows formed in water must be wrong. In other places, he found granites that were intruded into the preexisting layered secondary deposits of the flood, so not all granites were Primary rocks, cooled when the earth formed. For this Hutton and his followers were known as Plutonists, after the Roman name for the god of the underworld.

Hutton's ideas were not widely accepted at first, because his writing was rather convoluted and difficult to read, but the next generation of natural historians, especially former lawyer Charles Lyell, made Hutton's vision of an immensely old earth controlled by natural laws and processes (the concept of uniformitarianism, or the uniformity of natural laws in the past and the present) inescapable. Lyell's three-volume *Principles of Geology* (1831–1833) did such a good lawyerly job of proving the case for uniformity in geology that the Wernerian view was never again taken seriously.

Meanwhile, Lyell's contemporaries were finding it harder and harder to reconcile the increasing complexity of the geology of Europe with the simplistic notion that it was all deposited in a single Noah's flood. Between 1812 and 1826, the great naturalist Baron Georges Cuvier tried to sidestep the problem by claiming that the beds with the great prehistoric beasts (including the first marine reptiles found, like ichthyosaurs and plesiosaurs) were deposits of an "Antediluvian" period (before the flood) not mentioned in the Bible. In 1842, geologists such as Alcide d'Orbigny postulated twenty-seven separate creation events and floods not mentioned in the Bible to explain the long sequence of different fossils in different layers. By the 1830s and 1840s, even devout geologists like the Reverend William Buckland, a minister and natural historian at Oxford, gave up on their efforts to make the real rock record fit the Noah's flood story.

The Noah's flood model predicted a simple layer-cake worldwide sequence of coarse flood gravels and sands, overlain by a worldwide mudstone deposit. By contrast, the real geologic record is highly complex and variable from region to region, with intertonguing contacts between units and facies that change dramatically over relatively short distances within the same part of the sequence. It is full of thousands of individual mud-cracked layers and many different layers of salt and gypsum that simply cannot be explained by a single flood. There are thousands of different layers with delicate fossils in life positions, undisturbed by a single flood event.

By the late 1840s, the idea of Noah's flood was no longer relevant to explaining the real record of rocks on earth, and it was totally abandoned by devout Christians who did not doubt the Bible but no longer took it as a literal guide to nature. This was at least ten to twenty years before Darwin published his ideas on evolution. The old creationist accusation that scientists arrange the fossils in a sequence that supports evolution and then point to that same sequence as proof of evolution (supposedly a circular argument) is a laughably bad lie about history. The real sequence of fossils through time was discovered and documented by Christian scholars who wanted to reconcile the idea with Noah's flood but found that impossible.

Flood Geology Returns

In 1859, Charles Darwin published *On the Origin of Species*, and both the science of biology and our views about the world were forever changed. Controversial at first, as the years passed and each new edition came out (there were six in Darwin's lifetime), the evidence that life had evolved was so strong that by the time Darwin died in 1882, it was accepted by nearly all educated people in Europe and the Americas. Even in the American South, most educated people had no problem with evolution by 1900. For example, in 1880, the editor of one American religious weekly estimated that "perhaps a quarter, perhaps a half of the educated ministers in our leading Evangelical denominations" believed "that the story of the creation and the fall of man, told in Genesis, is no more the record of actual occurrences than is the parable of the Prodigal Son."[2]

This all changed in the early twentieth century with the birth of a movement called fundamentalism, a rejection of the modern Higher Criticism of the Bible. In a series of pamphlets published between 1895 and 1910, various authors asserted the "fundamentals" of the Protestant faith: the miracles of Jesus, his virgin birth, his bodily resurrection, his death on the cross to atone for our sins, and finally, that the Bible is the directly inspired word of God. Yet the early fundamentalists did not oppose evolution, since it was such a clearly established scientific fact, accepted by ministers as well as scientists. A. C. Dixon, the first editor of *The Fundamentals*, wrote that he felt "a repugnance to the idea than an ape or an orang-outang was my ancestor" but was willing "to accept the humiliating fact, if proved."[3] Reuben A. Torrey, who edited the last two volumes of *The Fundamentals*, acknowledged "for purely scientific reasons" that a man could "believe thoroughly in the absolute infallibility of the Bible and still be an evolutionist of a certain type."[4] Although the early fundamentalists were not happy with evolution, they were willing to live with it; they were not as stridently opposed to the idea as they would be a generation later. More importantly, evolution was accepted by most of the science textbooks of the time, so even if the parents were fundamentalists who rejected evolution, their children accepted it. Even in the conservative Baptist South, evolution was taught without much resistance in many educational institutions.[5]

Then came the turmoil of the Great War from 1914 to 1918, the progressive politics of Theodore Roosevelt and Woodrow Wilson, and the Great Influenza Epidemic of 1918. When the Roaring Twenties came along, America experienced a conservative backlash to all the changes and turmoil of the 1910s, and the country elected Warren Harding as president on the campaign promise of "a return to normalcy." Even as they voted for Prohibition of alcohol (and made a bunch of gangsters very rich, as everyone took their drinking private), the political winds also established a number of very regressive laws. The backlash against evolution led to several Southern states passing "monkey laws," which banned the teaching of evolution. This precipitated the famous Scopes Monkey Trial of 1925, a publicity stunt by the town of Dayton, Tennessee, and an attempt to legally test Tennessee's monkey law. The case was most famous for the battle between fundamentalist and three-time Democratic presidential candidate William Jennings Bryan and Scopes's attorney, Clarence Darrow. Bryan was humiliated on the witness stand for defending the absurdities of biblical literalism, and the media portrayed the trial as a victory of reason and science over benighted religious ignorance. But as the trial itself was inconclusive because of a mistake by the judge, the case was never appealed to a higher court, and many of those laws remained on the books for forty-three more years until 1967, when the Supreme Court struck them down in the case of *Epperson vs. Arkansas*.

Along with the view that evolution was false and the Bible was literally true, fundamentalism also revived the literal belief in Noah's flood. Even though almost a century earlier, devout Christian geologists had already rejected the idea that Noah's flood explained the rock record and the sequence of fossils through time, fundamentalists ignored the lessons of the past and tried once again to shoehorn the complexity of the geologic record into a single supernatural flood event. The first detailed attempt at such an explanation came from a Seventh-Day Adventist schoolteacher named George Macready Price, who published a series of books starting in 1902.

Price had no formal training or experience in geology or paleontology, and in fact, he attended only a few college classes at a tiny Adventist college. But inspired by Ellen G. White, the prophetess and founder of the Seventh-Day Adventist movement, he dreamed up an explanation called flood geology and aggressively promoted it for more than sixty years until his death in 1963. According to Price, the flood accounted for all of the fossil record; the helpless invertebrates were buried first, and the larger land animals floated to the top to be buried in higher strata or fled the floodwaters to higher ground. Price also originated the falsehood we just debunked about geologists dating rocks by their fossil content while simultaneously determining the age of fossils by their position in the geological column.

In Price's later years, his bizarre ideas about geology were generally ignored as embarrassments by most creationists.[6] Most subscribed to the day-age idea of Genesis, where the days of scripture were geologic ages of unspecified duration, and did not try to contort all the evidence of geology into a simplistic flood model. Some of Price's disciples actually tried to test his ideas and looked at the rocks for themselves, which Price apparently never bothered to do. In 1938, Price's follower Harold W. Clark "at the invitation of one of his students visited the oil fields of Oklahoma and northern Texas and saw with his own eye why geologists believed as they did. Observations of deep drilling and conversations with practical geologists [none of whom were trying to prove evolution but simply using biostratigraphy to find oil] gave him a 'real shock' that permanently erased any confidence in Price's vision of a topsy-turvy fossil record."[7]

Clark wrote to Price,

> The rocks do lie in a much more definite sequence than we have ever allowed. The statements made in the New Geology [Price's term for flood geology] do not harmonize with the conditions in the field. . . . All over the Middle West the rocks lie in great sheets extending over hundreds of miles, in regular order. Thousands of well cores prove this. In East Texas alone are 25,000 deep wells. Probably well over 100,000 wells in the Midwest give data that have been studied and correlated. The science has become a very exact one, and millions of dollars are spent in drilling, with the paleontological findings of the company geologists taken as the basis for the work. The sequence of microscopic fossils in the strata is very remarkably uniform. . . . The same sequence is found in America, Europe, and anywhere that detailed studies have been made. This oil geology has opened up the depths of the earth in a way that we never dreamed of twenty years ago.[8]

Clark's statement is a classic example of a reality check shattering the fantasy world of the flood geologists. Unfortunately, most creationists do not seek scientific reality. They prefer to speculate from their armchairs and read simplified popular books about fossils and rocks rather than go out in the field and do the research themselves or do the hard work of getting the necessary advanced training in geology and paleontology.

In the 1950s, a young seminarian named John C. Whitcomb tried to revive Price's ideas yet again. When Douglas Block, a devout and sympathetic friend with geological training, reviewed Whitcomb's manuscript, he

> found Price's recycled arguments almost more than he could stomach. "It would seem," wrote the upset geologist, "that somewhere along the line there would have been a genuinely well-trained geologist who would have seen the implications of flood-geology, and, if tenable, would have worked

them into a reasonable system that was positive rather than negative in character." He assured Whitcomb that he and his colleagues at Wheaton [College, an evangelical school] were not ignoring Price. In fact, they required every geology student to read at least one of his books, and they repeatedly tested his ideas in seminars and in the field. By the time Block finished Whitcomb's manuscript, he had grown so agitated he offered to drive down to instruct Whitcomb on the basics of historical geology.[9]

In 1961, Whitcomb and hydraulic engineer Henry Morris published *The Genesis Flood*, where they revived Price's notions with a little twist or two of their own. Their main contribution was the idea of hydraulic sorting by Noah's flood, where the flood would bury the heavier shells of marine invertebrates and fishes in the lower levels, followed by more advanced animals such as amphibians, reptiles (including dinosaurs) fleeing to intermediate levels; finally, the "smart mammals" would climb to the highest levels to escape the rising floodwaters before they were buried.

Henry Morris championed flood geology for the rest of his life (he died in 2006), and his biases about science and creationism were quite clear in his writings. For example, he wrote, "The main reason for insisting on the universal Flood as a fact of history and as the primary vehicle for geological interpretation is that God's Word plainly teaches it! No geological difficulties, real or imagined, can be allowed to take precedence over the clear statements and necessary inferences of Scripture."[10]

Compare this attitude with our description of the scientific method we discussed in chapter 1. Unlike real scientists, who follow the evidence wherever it leads and must be ready to reject even their most cherished hypotheses if the data require it, creationists work from assuming their conclusions to be true (the Bible is a literal description of history) then bending and twisting and ignoring facts that don't fit it. This is why creationism is not and can never be scientific, even if creationists appropriate that title when they are trying to fool the courts and public by calling themselves "scientific creationists."

How Do We Know?

As in other chapters, it is worthwhile to give the detailed scientific evidence for why scientists reject a specific weird idea about the earth—in this case, flood geology.

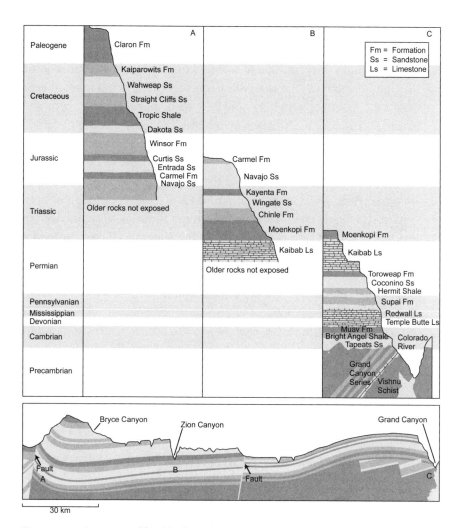

Figure 10.1. Sequence of fossil beds in the Colorado Plateau region, showing the record that spans from the Cambrian to Permian in the Grand Canyon, the Mesozoic in the Zion Canyon area, and the Cenozoic in Bryce Canyon. (*From Prothero, 2013.*)

1. As soon as any advanced geology student who has real field experience reads this bizarre notion of smarter animals outrunning the dumber ones, we are astounded at its silliness and naivete. Whitcomb and Morris's explanation of fossil sequences is based on oversimplified diagrams from children's books, which show invertebrates at the bottom, dinosaurs in the middle, and mammals on top. Their idea does not correspond to the actual reality of the rock record (as people like the Reverend William Buckland knew in the 1840s). For example, most of the marine invertebrates are still with us and are often found in fossil beds *above* the mammal bones, so how does their model explain clams and snails outrunning the smarter mammals? In places like the Colorado Plateau (fig. 10.1), we can see the actual sequence of rocks from the Grand Canyon, which covers much of the Paleozoic Era; to Zion National Park, which covers most of the Mesozoic Era; to Bryce Canyon National Park, which is built of pinnacles of Eocene limestones. Even within this nearly complete sequence from Cambrian to recent, we can easily falsify the Genesis flood model. The Colorado Plateau sequence yields "dumb" marine ammonites, clams, and snails from the Cretaceous Mancos Shale *on top of* "smarter, faster" amphibians and reptiles (including dinosaurs) from the Triassic and Jurassic Moenkopi, Chinle, Kayenta, and Navajo formations.

2. Just to the north, in the Utah-Wyoming border region, the middle Eocene Green River Shale yields famous fish fossils that have been quarried by commercial collectors for almost a century. The Green River Shale produces fossils of freshwater fish as well as freshwater clams and snails, frogs, crocodiles, birds, and land plants. The rocks are finely laminated shale which clearly was deposited in quiet water lakes over thousands of years, with one layer after another of fossil mudcracks and salts formed by complete evaporation of the water. These fossils and sediments are all characteristic of a lake deposit that occasionally dried up, not a giant flood. These Green River fish fossils lie *above* the famous dinosaur-bearing beds of the upper Jurassic Morrison Formation in places such as Dinosaur National Monument and above many of the mammal-bearing beds of the lower Eocene Wasatch Formation as well, so once again the fish and invertebrates are found above the supposedly smarter and faster dinosaurs and mammals.

3. If you think hard about it, why should we expect that marine invertebrates or fish would drown at all? They are, after all, adapted to marine waters, and many are highly mobile when sediment is shifting. As Stephen Jay Gould put it, "Surely, somewhere, at least one courageous trilobite would have paddled on valiantly (as its colleagues succumbed) and won a place in the upper strata. Surely, on some primordial beach,

a man would have suffered a heart attack and been washed into the lower strata before intelligence had a chance to plot a temporary escape. . . . No trilobite lies in the upper strata because they all perished 225 million years ago. No man keeps lithified company with a dinosaur, because we were still 60 million years in the future when the last dinosaur perished."[11]

4. In addition to the examples just given, there are hundreds of other places in the world where the "dumb invertebrates" that supposedly drowned in the initial stages of the rising flood are found on top of "smarter, faster land animals," including many places on the Atlantic Coast of the United States, in Europe, and in Asia, where marine shell beds overlie those bearing land mammals. In some places, like the Calvert Cliffs of the Chesapeake Bay in Maryland or Sharktooth Hill near Bakersfield, California, land mammal fossils and marine shells are all mixed together, and there are also beds with marine shells *above and below* those containing land mammals! How could that make any sense with the rising floodwaters of the creationist model?

5. I did my graduate dissertation work in the Big Badlands of South Dakota, one of the richest vertebrate-bearing fossil deposits in the world. The sequence of fossils there is very well known, and we can now establish the precise ranges of species through a thickness of several hundred feet of sandstones and mudstones (fig. 10.2). At the base of the sequence are marine fossils, but right above them are the late Eocene fossils of the Chadron Formation, which include many large and spectacular mammals, including the huge rhino-like brontotheres. Above these in the overlying Brule Formation is a different assemblage of fossil mammals, none of whom look like they could have outrun the huge brontotheres. Many of these are rodents. It's hard to imagine them doing a better job at scrambling for higher ground than the bigger, longer-legged animals. The clincher, however, is the fact that the most abundant fossils in the Brule Formation are tortoises! We have a new version of Aesop's fable of the tortoise and the hare, although here the dumb tortoises beat not only the hares to higher ground but also nearly all the rest of the smarter, larger, longer-legged mammals as well. If there ever was a clear-cut falsification of the flood geology model, this alone should be enough!

6. The creationists' favorite example is always the Grand Canyon (fig. 10.1). A spectacular sight that draws millions of visitors from all over

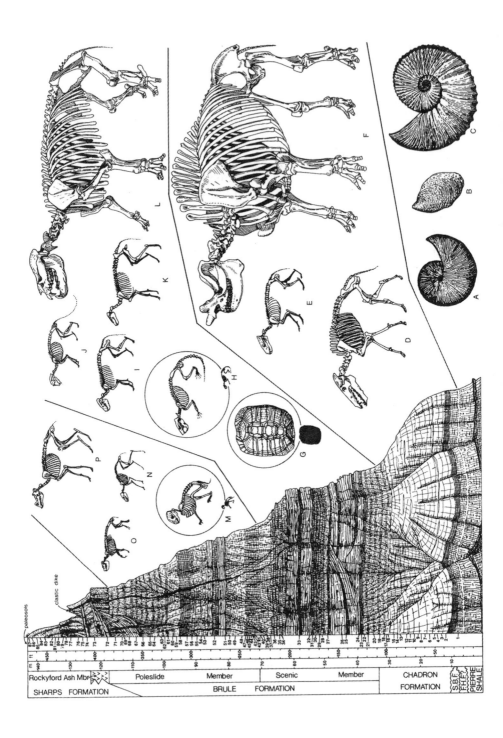

paleosols

clastic dike

| Rockyford Ash Mbr | Poleslide | Member | Scenic | Member | CHADRON |
| SHARPS FORMATION | BRULE | FORMATION | | | FORMATION |

S.B.F.

PIERRE SHALE

Facing, Figure 10.2. The sequence of fossils in the Big Badlands of South Dakota, showing huge long-legged animals like the horned brontotheres and other fast mammals from the Chadron Formation that apparently could not outrun a land tortoise to higher ground. (*Courtesy G. J. Retallack.*)

the world, it attracts the attention of creationists because it is one of the few places on the planet where the layers seem relatively simply deposited and flat lying, so they can be shoehorned into a flood geology model. Creationists have even managed to get one of their books on the subject sold in the bookstore of the South Rim Visitors Center, although I'm told it's not in the science section but in the religion section. I've hiked the Grand Canyon dozens of times, rafted it twice, and flown over it several times, so I've seen it from top to bottom. Depending on which version you read of the creationists' Grand Canyon story, the entire Grand Canyon was deposited in a single flood event described by Genesis 6–7 and then eroded quickly when the floodwaters drained away and carved it.

Let's examine the problem from a scientific perspective. Any sedimentary geologist (as I am) knows that actual flood deposits start with a layer of pebble and cobble conglomerate from the initial rush of high-energy floodwaters, then possibly a layer of sand as the water slows down, and then thick deposits of mud, becoming mudstone or shale—and that is all.

How does this compare to the real Grand Canyon? It's not even close! Even a cursory glimpse at the sequence of layers in the Grand Canyon (fig. 10.1) shows that it is highly complex and cannot be explained by a single superflood (or even many floods, if that were an option). For one thing, there is no great deposit of coarse gravel, boulders, and sand near the base representing the high-energy phase of rapidly moving water. For another, the upper part of the Grand Canyon sequence above is not just a single thin layer of mud. Instead, it is a complex sequence of shales (*not* mudstones), sandstones, and limestones that alternate in a form that resembles no known flood deposits.

Let's start at the very bottom of the canyon. Instead of coarse gravel, sand, and boulder deposits that flood geologists might expect, we have the ancient rocks of the Unkar and Chuar groups (fig. 10.1). These are mostly quiet-water shales, plus sandstones and even some limestones.

Many of these limestones contain stromatolites (fig. 10.3A), dome-like mounds of layered sediment formed by algal mats that can only grow in the quiet waters of a sunny coastal lagoon. The individual layers in these stromatolites testify to hundreds of years of growth on each one, and there are multiple layers of stromatolites, each representing a separate episode of slow growth followed by burial and then another phase of growth on a new surface. And this was supposedly formed during a single huge flood event only forty days in duration?

A clear falsification of the flood geology model is the abundant mudcracks (fig. 10.3B) found in many of the shale units of the Unkar and Chuar groups. We've all seen mud dry up and form cracks, and common sense should tell even the creationists that the entire muddy surface was deposited and then dried up, not formed during the inundation of a flood. There's not just one layer of mudcracks but hundreds of them, sometimes stacked in a long sequence. Clearly, these rocks represent dozens of small episodes of mud deposition and then complete drying, not a single catastrophic flood.

Even more strongly falsifying the flood geology model is that in the middle of this Unkar-Chuar sedimentary sequence are the Cardenas lava flows, dozens of individual flows totaling almost 300 meters (1,000 feet) in thickness. If these rocks had erupted into the floodwaters, they would be entirely composed of blobs of lava known as pillow lavas, which we can see erupting from undersea lava flows today. Instead, the Cardenas lavas show clear signs that they are normal subaerial eruptions and flowed downhill from their nearest volcano, not unlike the lavas erupting from Mount Kilauea in Hawaii. The very top of the lava flows show evidence that they had completely cooled and were even weathered and eroded by wind and rain before the next sequence of sedimentary rocks was deposited on top of them. This is hardly consistent with the idea of lavas erupted underwater during a major flood!

Finally, the clincher is that all these ancient Unkar and Chuar rocks at the base of the Grand Canyon are now found tilted on their sides, their edges eroded, and then the rest of the Grand Canyon sequence is deposited on top of them (fig. 10.1). How the heck does a flood geologist explain this? If these rocks were all soft soupy sediments deposited by Noah's flood, then as soon as some supernatural force rapidly tilted them on their sides, they would have all slumped downhill and left big gravity slump folds, a feature well known to sedimentologists. Instead, the entire sequence is undisturbed and full of stromatolites, mudcracks, and lava flows that belie the entire flood geology model right then and there. We have evidence of deposition of the Unkar and Chuar sediments (along with long erosion between them, when the Galeros lavas flowed across the landscape), then the hardening of these soft sediments into sedimentary rock layers, then tilting, then erosion, and then *another* long sequence that makes up the upper part of the rocks of the Grand Canyon. All of this is supposedly formed in a single large flood event?

And so it goes, layer by layer, right up through the rest of the Grand Canyon. The first unit above the tilted Unkar and Chuar rocks is the Tapeats Sandstone (figs. 10.1, 10.3C), a classic beach and nearshore deposit. It is chock-full of trackways and burrows of trilobites, worms, and other invertebrates, layer after layer. When would these animals have had any time to crawl across the bottom and leave tracks or to burrow through the sediment if it had been rapidly dumped by a flood? Above the Tapeats is the Bright Angel Shale, which geologists interpret as deposited on a shallow marine shelf below the action of storm waves. It, too, is full of tracks and burrows, but of the types that today occur in the deeper part of the ocean. How did these tracks and burrows get there,

Following pages, Figure 10.3. Close examination of the actual rocks in the Grand Canyon makes the flood geology hypothesis completely absurd. A. Large mudcracks in the Precambrian Grand Canyon Series, in the lowest tilted sequence in the Grand Canyon. There are layer after layer of cracks like these in these shales, showing that there were hundreds of individual drying events—not possible with a single flood. (Photo by the author.) B. In other places, there are layered algal mats known as stromatolites, which were formed by daily fluctuations of sediment and algal growth. Some actually record decades or centuries of growth. These are abundant in the tilted late Precambrian limestones beneath the Paleozoic rocks of the Grand Canyon. (Photo courtesy US Geological Survey.) C. The lower Cambrian (*left*) Tapeats Sandstone and (*right*) Bright Angel Shale are full of layer after layer of sediments with complex burrows and trackways, showing that each layer had been part of another sea bottom that was crawled upon and burrowed into and then buried again and again. (Photo courtesy US Geological Survey.) D. The Pennsylvanian-Permian Supai Group and Hermit Shale are also full of layer upon layer of mudcracks, showing that they went through hundreds of episodes of drying, completely falsifying the flood geology model. (Photo courtesy US Geological Survey.) E. The Permian Coconino Sandstone is composed entirely of huge cross-beds that could only have formed in desert sand dunes, not underwater. (Photo courtesy US Geological Survey.) F. The Coconino dune faces also are covered with the trackways of reptiles that could never have been formed underwater. (*Photo courtesy US Geological Survey.*)

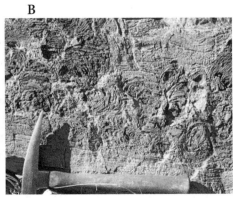

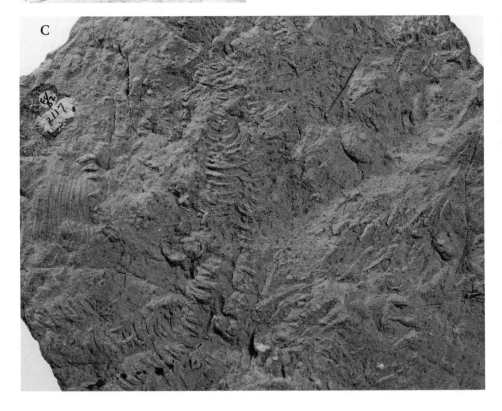

D

E

F

layer after layer, if all the deposits of the Grand Canyon are a single flood deposit that drowned and buried all the marine life before they had a chance to begin burrowing?

The Bright Angel Shale has a complex interfingering relationship with the next unit above, the Muav Limestone. These types of relationships, where a thin layer of limestone alternates with a thin layer of shale, are very typical of deposits we find today when sea level slowly fluctuates back and forth, but it is impossible to explain such a complex relationship by a single flood dumping these sediments in a flat layer cake. The Muav Limestone is one of three consecutive limestones forming the steepest cliffs in the Grand Canyon. Above the Muav is a sharp erosional surface with deeply eroded collapse features (from ancient collapsed caves slowly dissolved out of the Muav), into which the much younger Temple Butte Limestone is deposited. The Temple Butte was then eroded away in most places except for the remnant fillings of those collapse features, and above it is deposited the big cliff of the Redwall Limestone.

All three limestones have the features typical of modern limestones, made largely of the delicate remains of fossils. Today, we find such sediments forming in tropical, clear-water lagoons or shallow seas, such as those in the Bahamas or Yucatan or the South Pacific. In no case do these sediments form where there is the huge energy of floods or lots of mud stirred up by floodwaters. Particularly diagnostic is the fact that many of the fossils are extremely delicate (such as the lacy moss animals, or bryozoans), yet they are intact and undisturbed, which proves that the flood cannot have occurred. Even more evocative are the delicate animals, such as the sea lilies (crinoids) and lamp shells (brachiopods), which are sitting just as they lived on the seafloor, layer after layer growing over and over, undisturbed by high-energy currents and buried by lime mud (not flood-type mud) that gently filtered in around them without disturbing them. This is true of limestones like this around the world, so the Grand Canyon is not a special case and definitely not evidence of a supernatural flood. The same is true of the Toroweap and Kaibab limestones, which form the rim of the Grand Canyon.

Above the Redwall Limestone are the alternating sandstones and shales of the Supai Group, followed by the red Hermit Shale. The sandstones of the Supai Group are full of small ripples and small crossbeds, features of gentle deposition in rivers, not raging floodwaters or the muds settling out after the flood movement had stopped. The Supai Group and Hermit Shale contain layer after layer of mudcracks (fig. 10.3D), clearly demonstrating repeated drying out, as well as delicate plant fossils preserved intact, which is hard to explain by energetic floodwaters. These plant fossils are of extinct groups not found in any of the layers below or above, so how did they get there?

Finally, the clincher is the distinctive white band that is visible just below the rim on both sides of the canyon, known as the Coconino Sandstone. This unit has huge cross-beds (fig. 10.3E) that are only known to form in large-scale desert sand dunes, *not* underwater. They also have small pits characteristic of the impacts of raindrops. How did raindrops land on these surfaces if they were immersed in a great flood? Even stronger proof is that many of the dune surfaces are covered with trackways of land reptiles (fig. 10.3F). How did these dry sand-dune features and dry-land reptile trackways form under a huge flood event? I've read the creationists' attempts to explain these features, and they are classic examples of special pleading and twisting and distorting scientific evidence as they thrash around with their completely unconvincing scenarios.

The most impossible thing the creationists ask you to believe is this: the entire pile of sediments of the Grand Canyon sequence, soft and soupy and supposedly deposited during a single flood event, was then eroded down to form the present-day Grand Canyon by the recession of the floodwaters. Wait a minute—didn't they just use the recession of the floodwaters and the settling out of still water to deposit the thick piles of postflood shales, sandstones, and limestones in the first place? Or if that's not their scenario, then how did a soft pile of wet mud, sand, and lime hold up without slumping down and sliding into the gorge as the torrential retreating floodwaters rushed through? Did the creationists also suspend the laws of gravity? Anyone with common sense can watch the Grand Canyon as it erodes today, with the long-hardened sediments (now sedimentary rocks) slowly weathering and eroding, dropping into the canyon by the action of gravity or by rains and small local canyon floods, and then being slowly carried away by the erosion of the Colorado River. There's no need for the bizarre flood scenario, unless your religious blinders are so dense that you cannot tell common sense from fantasy anymore.

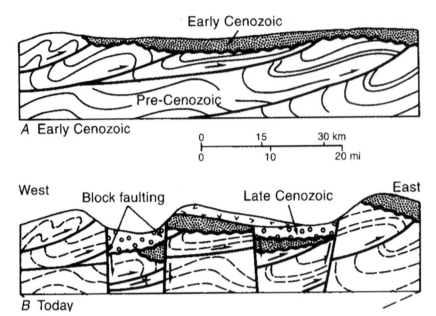

Figure 10.4. Creationists try to align relatively simple flat-lying sequences like the Grand Canyon with Noah's flood and ignore the vast majority of geologic settings around the world that in no way resemble a layer cake that could be deposited by a flood. For example, in the Basin and Range Province of Utah and Nevada, just north of the Grand Canyon, the geologic relationships are extremely complex. A. Paleozoic and Mesozoic beds are faulted and folded many times, with Paleozoic beds full of marine shells overthrust above Mesozoic dinosaur-bearing beds (contrary to the idea that dinosaurs could outrun the marine invertebrates in the rising flood). These older beds are then eroded off and unconformably overlain with early Cenozoic beds containing fossil mammals. B. The early Cenozoic beds were then cut by Miocene normal faults, and the basins were filled with late Cenozoic sediments containing extinct horses, camels, mastodons, and other Miocene land mammals. None of this complex geometry could be explained by simplistic Noah's flood models. (*Modified from Prothero, 2013.*)

7. More important than the details of the Grand Canyon sequence is the fact that 99 percent of the rocks on the planet look nothing like the layer-cake sequence of the Paleozoic rocks in the Grand Canyon. Anywhere else you choose to look on earth, the geology is much more complex and cannot possibly be explained by a single Noah's flood. For example, just a few hundred kilometers away from the canyon are typical exposures of rocks in Nevada and Utah (fig. 10.4). These are incredibly complicated, with layered sedimentary rocks folded and faulted and eroded off, then

buried by younger sedimentary rocks that are also folded and faulted in much later events. So which rocks were formed during the flood, and how did they all form in one event, when each sequence of folded sedimentary rocks is deformed and faulted by younger events? Creationists make no effort to address this problem but instead lead their flocks to the Grand Canyon again and again, repeat their long-debunked stories from the flood geology models of Whitcomb and Morris and their successors, and never learn anything new.

8. The most significant implication of flood geology and its fantasy view of the earth is a practical problem. Without real geologists doing their work, none of us would have the oil, coal, gas, groundwater, uranium, and most other natural resources that we extract from the earth. There are lots of devout Christians in oil and coal companies (I know many of them personally), but they all laugh at the idea of flood geology and would never attempt to use it to find what they're paid to find. Instead, they have seen the complexity of real geology in hundreds of drill cores spanning whole continents and don't even begin to try to interpret these rocks in a creationist mold, even though they may be devout Christians and believe much of the rest of the fundamentalist's credo. If they tried, they'd find no oil and would lose their jobs! As creationists keep trying to get their bizarre notion of flood geology inserted into classrooms and places like the Grand Canyon, we have to ask ourselves, are we willing to give up the oil, gas, coal, groundwater, and uranium that our civilization requires? That would be one of the steepest prices we would pay if we followed the creationists.

A Surreal Journey with Creationists

In 2012, I got to experience the backward looking-glass world of creationists firsthand. I was invited to be a guest scientist on the British reality TV series *Conspiracy Road Trip*. The premise of the series is that the host (Andrew Maxwell, an Irish comedian) travels with five young believers of some crazy idea, taking them to key locations and putting them in front of evidence that challenges their beliefs. They had already done episodes on UFO believers, the 7/7 bombings in London, and 9/11 Truthers, so the next group of crazies in line was the creationists. The producer explained that he wanted me and a number of other scientists to meet at important locations (I was to film on the rim of the Grand Canyon), show these creationists the actual scientific evidence, and let them squirm and try to rationalize their beliefs in front of the cameras. Even though I've battled creationists in debates and TV panels before and have done TV documentaries in the field on prehistoric animals, I'd never done something that combined the two. I've written about and argued enough with creationists to know them and their arguments (and the scientific reality) down pat. Still, I prepared for anything. I even brought along a bunch of real fossils to pass around, and I put my key diagrams on a series of huge laminated flip charts.

In mid-April 2012, they flew me out to Las Vegas, where I went out to a late dinner with the producer, who looked over my materials and walked me through his plans for filming. He wanted to hold off my confrontation with the creationists until the cameras were rolling, so he asked me to try to avoid them at breakfast in our hotel that morning. This was nearly impossible since we were the only group at breakfast, and later, one of them recognized me while we waited in the airport. Then we flew out of the Henderson, Nevada, airport on small prop planes to see the entire Grand Canyon from the air (an amazing flight that I had never done before). The five creationists and Andrew Maxwell, plus the director and two cameramen were in one plane, while the producer and a camerawoman were in a smaller plane with me. After landing on the South Rim airport, we rode in a minibus to lunch at Grand Canyon Village (where they tried to chat to me again), then we all went out to Lipan Point on the South Rim, one of the best places for an unobstructed view of the eastern Grand Canyon with no fences and fewer crowds and less background noise. After they got a chance to look out over the rim, the film crew set me up with my back to the canyon and began our first segment.

As I began my explanation (I come in at about 7:50, and you can find the video link by searching "conspiracy road trip creationism" on your browser),[12] I tried to be very straightforward and clear about the rules of science and about why science must reject supernatural events as untestable, why the principles of geology require huge amounts of time to explain things like the great angular unconformity at the base of the Grand Canyon (which cannot be explained by a single flood event), and why each layer has clues in it that refute the Noah's flood model. I even pulled out of my backpack a small collection of representative fossils to show that the Grand Canyon records the changes in fossils through time. I pointed out that although flood geology was widely believed before 1800, by 1840, devout British creationist geologists themselves rejected it because the rock record clearly does *not* support the notion of Noah's flood.

Only one of the five, a guy named Phil (no last names were given), was familiar with all the standard flood geology interpretations of the Grand Canyon by the likes of Steven Austin and John Woodmorappe. (I later learned his name is Phil Robinson. He has no background in science or geology but is a physical education teacher. He is also chair of the Northern Ireland Creation Outreach Ministries.) Naturally he tried to argue with me about arcane points that only he and I knew about, which drove the producer and director crazy. The only other one to speak out was a tall Muslim creationist named Abdul, who was very loud and assertive and spent most of his time shouting down others and arguing that science is crap (not a very good way to convince people). The two women and the other guy were almost silent the whole day, with nothing much to contribute. We filmed at Lipan Point for better than an hour before the producer and director had had enough of our arguing in circles. Then we hopped into our vehicles and headed for our next stop.

For the rest of our day in the gorgeous scenery of the region, I had originally thought about giving them a broader overview of the wonders of the geology of the Colorado Plateau from places like the Echo Cliffs monocline or the Navajo Sandstone dune cross-beds (as I had long done for my college geology classes). After our first encounter, however, I could see that it was pointless. Instead, the producer had the clever idea to take them to Horseshoe Bend (fig. 10.5), just south of our hotel in Page, Arizona. Here one can see an example of huge river meanders (normally formed near the mouth of a river, where the gradient is very low and the river cuts sideways rather than downward). These meanders are then incised into deep canyons like the Grand Canyon or, an even better example, the Goosenecks of the San Juan River in Utah.

Steep-walled canyons like these are exclusively formed by rapid uplift far above sea level, and to a geologist, they only make sense if the Colorado Plateau was once near sea level (as many lines of evidence now support) and then later uplifted to cause sea-level river meanders to carve down into hard bedrock. Whatever you think of these features, they are *not* consistent with the rapid draining of water from the earth's surface after Noah's flood, the mechanism that creationists claim cut the Grand Canyon. We set up a very simple demonstration where we simulated the draining of the floodwaters

Figure 10.5. The famous entrenched meanders of the Colorado River at Horseshoe Bend, just south of Page, Arizona. (*Courtesy Wikimedia Commons.*)

with a bucket of water running down a sandy slope. Although our sand substrate was pretty porous and most of the water soaked in, you could still see the straight stream channels that form any time river waters are moving rapidly downslope during a flood. What's more, it was clear that there was no way such a flood could form the broad lazy meander beds we saw before us.

Sure enough, the producer was right: this demonstration was very effective and caught the smug Phil and others completely flat-footed, since it was simple laws of physics and geomorphology in action. As often happens, creationists were unable to answer this puzzle, since they have no real understanding or firsthand experience in geology; they simply have memorized ad hoc explanations for specific areas like the Grand Canyon. Rather than admitting they didn't have the answer, Phil argued that he was sure there *must be* a creationist answer to this puzzle. Nevertheless, Andrew Maxwell kept at them and made them confess that they couldn't deny basic physics.

I flew home from Page the next morning. For the rest of the trip, the creationists were confronted by a number of other scientists who effectively argued why the flood geology model and the Noah's ark story were scientifically impossible. The most effective person of all, however, was anthropologist Tim White at University of California, Berkeley (famous as the codescriber of the famous hominin fossil Lucy, properly known as *Australopithecus afarensis*, from Ethiopia). For his demonstration, he laid out replicas of a number of hominin skulls and had the creationists sort them by their anatomy. Once they had done so, he pointed out that this was the exact sequence that these skulls were found in a single place in Ethiopia and that primitive ones were never found on the level with the advanced ones, and vice versa. It was a remarkable bit of scientific theater, and they were unable to respond coherently to it since there *is* no creationist response. The most primitive skulls looked like "apes" to them, the most advanced ones were clearly "human," and there in front of them were all the intermediates in between.

But our scientific lessons ended up being a tiny part of the hour-long episode, which is largely filled with footage that one finds in many reality shows, from *Survivor* to MTV's *The Real World*: people are cooped up together on camera and begin to squabble among themselves. Most of the episode focused on the antics of the five creationists as they rode their bus through two thousand tedious miles of American interstates over an entire week, got into fights, split into factions, argued with each other, and generally acted immaturely and thoughtlessly. Occasionally, Andrew would get one to interview with him directly. Although the blonde girl, JoJo, seemed to be changing her mind, the rest were still dogmatic and inflexible. (Amusingly, Abdul was completely unaware of how much he had alienated everyone, failed to make good arguments, and believed that Islam had triumphed over Christianity and science in this exercise.) None of them could give coherent answers to the scientific evidence; nonetheless, they were determined to stick with their beliefs.

This was no surprise to any of us, since evidence doesn't matter to creationists. They have an entire worldview that is wrapped around the salvation of their immortal soul and the fear of rejecting the literal interpretation of the Bible (or, in the Muslim case, the Quran). That comes first, and everything else is unimportant. They reject evolution and modern geology only because they've been told to do so by religious leaders, even though they have no clue what it's about; what they *think* they know about it is wrong. Indeed, they showed the classic response of true believers: when something threatens your worldview, you cling to it even more strongly and find any way you can to dismiss or ignore contrary evidence. That, apparently, is the point of the entire show, since the 9/11 Truthers and the UFO believers acted the same way. But given the way the show was framed, it's clear that the producers want to put these creationists on camera as object lessons on how irrational, dogmatic, and impervious to evidence they really are, while showing less dogmatic viewers that scientists can be friendly and reasonable and can have all the evidence. Given the low level of creationist beliefs in the United Kingdom, this is probably not a hard sell, but I am not sure whether it ever aired in the United States, where creationism still claims about 40 percent of the population.

The topic of flood geology has been debunked in a number of places, so I won't spend more time discussing it. The evidence given already is sufficient to consign it to the garbage bin of debunked ideas, just as devout geologists of the 1840s like Reverend William Buckland were honest enough to admit. Let's move on to something else.

⑪

Are Dinosaurs Faked?

Fake Deniers Claiming Dinosaurs Are Faked

The internet is notorious as a cesspool of lies and misinformation, but once in a while you run into something so outrageous that you have to ask yourself, is this real or is someone trying to troll you for a reaction?

In 2015, a website for a group calling itself Christians Against Dinosaurs (CAD) went viral, and at least a dozen of my Facebook friends forwarded it to me in surprise, curious as to whether I'd seen it and if it was real.[1] Christians Against Dinosaurs managed to get covered in the *Huffington Post* and in several other online media outlets, giving it even more exposure.[2] The site claims that dinosaurs (and fossils in general) are all a big lie to undermine the Christian faith and that fossils seen in museums are faked or sculpted out of rock by Big Paleo trying to make millions by fraud. Prominently featured on the site, on Facebook, and on YouTube are short videos by a young woman who makes these very claims, arguing that paleontologists fabricate fossils out of rock to look like animals.[3] Even more shocking and hilarious, she seems to think that such faked fossils are worth millions of dollars and keep Big Paleo afloat. As the *Huffington Post* described it,

> "A fossil is not actually a piece of bone," she says in the video. "It's actually a bone that was once in the ground that has been filled with limestone, calcium, and other stone-like deposits, so at the end of the day, it's a rock made out of rocks. So, you have a rock that's [six inches long] and you hand it to a paleontologist, who chips away at it until you have something looking like a bone—and that is a fossil," she continues. After covering a table with broken pieces of . . . something, she tells viewers to pretend they are paleontologists (ooh, activity time!) and put the shards back together in their original form. She even offered some spackle to assist in the reconstruction. But, she says, it's supposed to be a brachiosaurus skull—and "If

you're a paleontologist and you want to keep your job, you turn that into a brachiosaurus skull."[4]

The CAD people were thrown off the parenting forum Mumsnet after the following post was prominently featured:

I'm really concerned about dinosaurs, and I think something needs to be done. The science behind them is pretty flimsy, and I for one do not want my children being taught lies. Did you know that nobody had even heard of dinosaurs before the 1800s, when they were invented by curio-hungry Victorians?

Charles Darwin's later theory of evolution entirely disproved dinosaurs, yet the dinosaur lie was twisted and adapted to try to make it fit. Any proper look at the facts will reveal that dinosaurs simply never existed.

Aside from the educational aspect, dinosaurs are a very bad example for children. At my children's school, several children were left in tears after one of their classmates (who had evidently been exposed to dinosaurs), became bestially-minded and ran around the classroom roaring and pretending to be a dinosaur. Then he bit three children on the face. One poor girl has been left with a severely dented nose and the whole class was left traumatised by this horrible display.

Nothing about dinosaurs is suitable for children, from their total lack of family values through to their non-existence from any serious scientific point of view.

Recently my sister foolishly gave my two youngest some dinosaurs toys for Christmas. After telling her to get out of my house I burnt the dinosaurs. My children were delighted because they know that dinosaurs are evil. I am fortunate that my family has been very supportive, and has disowned my children's former aunt.

Please, do what you can to get dinosaurs taken off the curriculum. Our school has been recently presented with a 214-signature petition, and following that and our recent protest the headmaster has said that he will take it to the governors. We are lucky that he is so sympathetic to our cause, but I fear that others may not be.

If you would like to lend your support to our campaign, we have a Facebook group where we spread facts and research about the dinosaur myth. Hope to see you there!:-)[5]

This seems so exaggerated that it immediately struck me as a likely example of extreme satire and parody, which are so common on the internet. Often referred to as a *Poe*, these satirical pieces are intended to mock the bizarre beliefs of many groups of people from the extreme political and religious fringes. The idea was first coined by Nathan Poe in a 2005 post,[6] and Poe's law is the "observation that it's difficult, if not impossible, to distinguish between parodies of fundamentalism or other extreme views and their genuine proponents, since they both seem equally insane."[7] How can you create a parody or mockery of people who seriously believe that the earth is the center of the universe, such as the extremist Catholics who run the www.galileowaswrong.com site and hold geocentrism meetings (see chap. 3)?[8] Their thinking is so outrageous that it borders on self-parody, as do the views of Christians Against Dinosaurs.

This idea that fundamentalists reject dinosaurs as unbiblical even goes against recent trends. Most of the media-savvy creationists, from Ken Ham's Answers in Genesis organization to convicted felon Kent Hovind, who called himself "Dr. Dino" (even though his degree is from a diploma mill and he knows nothing about dinosaurs), have cashed in on the popularity of dinosaurs. They have embraced them as a way to lure in more followers, bending the words in Genesis in order to make dinosaurs fit somehow.

So is Christians Against Dinosaurs a well-disguised parody or the real thing? Several internet-savvy people commenting on my Facebook page dug up the source of the site and found that the admins are not fundamentalist Christian groups; instead, some of the admins have associations with other parody sites. The young woman with the two viral YouTube videos was eventually tracked down. Her name is Kristen Auclair, and at the time she worked in an insurance company in Massachusetts as one of the admins of the CAD site (another clue in the acronym?). The famous atheist activist Aron Ra, who has produced many YouTube videos challenging religion, got her to agree to an interview. In the interview,[9] Auclair stuck to her guns and answered him with a straight face; she seemed to be dead serious in denying dinosaurs ever existed. So maybe it's real? But then someone else did a bit more digging and found that in her yearbook,[10] Auclair was voted Class Clown! So maybe she's a really convincing actor and the whole thing is a bit of performance art? In late 2015, her charade was busted when others convincingly demonstrated that she was trolling the creationists and those of us who battle creationism on the internet.[11]

But even if she is a good faker, she has apparently attracted a lot of true believers to her site who agree with her in all seriousness and think dinosaurs are a challenge to their faith. The Christians Against Dinosaurs Facebook page is filled with comments by what appear to be people who sincerely believe that fossils are faked![12] As discussed below, there really *are* fundamentalists who preach that fossils (especially dinosaurs) are frauds perpetrated by scientists to make themselves rich and/or to push their evolutionary ideas on us and destroy our faith. All Kristen Auclair had to do was put up outrageous YouTube videos and create a site for a phony Christians Against Dinosaurs organization and lots of unsuspecting fundamentalists were sucked in.

Real Dino Deniers

But the sites for the genuine dinosaur deniers shocked even my sense of how low pseudoscientists and religious extremists can go. We are all familiar with how creationists use ad hoc explanations and special pleading to rescue the absurdities of their worldview, from trying to cram all the animals into Noah's ark and dismissing the huge numbers problem, through their nonbiological concept of created kinds, to doing all sorts of violence to the geologic record to justify the Noah's flood story (see chap. 10), to even insisting that men have one less rib than women do (the last one is easy to check; they don't). As I have discussed in previous chapters, the more extreme biblical literalists also believe in a flat earth and reject the heliocentric solar system. But I was flabbergasted to read of a whole group of extreme creationists who really do insist that dinosaurs are a hoax and never existed![13] With something as widely accepted, exciting, and popular as dinosaurs, which people can see for themselves in their local museum, how could any person in the twenty-first century argue they are not real?

Yet that is exactly the position of this odd creationist subcult, which would be unknown and invisible to most of us were it not for their web presence, whose web design is not as garishly bad as most fringe websites. You can scan their website, http://www.ocii.com/~dpwozney/dinosaurs.htm, and see just how far off the deep end they have plunged. Here is a representative quote for how these people argue that paleontologists are creating fraudulent dinosaur fossils:

What would be the motivation for such a deceptive endeavor? Obvious motivations include trying to prove evolution, trying to disprove or cast doubt on the Christian Bible and the existence of the Christian God, and trying to disprove the "young-earth theory." Yes, there are major political and religious ramifications.

The dinosaur concept could imply that if God exists, he may have tinkered with his idea of dinosaurs for awhile, then perhaps discarded or became tired of this creation and then went on to create man. The presented dinosaur historical timeline could suggest an imperfect God who came up with the idea of man as an afterthought, thus demoting the biblical idea that God created man in His own image. Dinosaurs are not mentioned in the Hebrew Bible.

Highly rewarding financial and economic benefits to museums, educational and research organizations, university departments of paleontology, discoverers and owners of dinosaur bones, and the book, television, movie and media industries may cause sufficient motivation for ridiculing of open questioning and for suppression of honest investigation. [That's a real laugh! Most paleontologists are poorly paid, and few even get a job in paleontology!][14]

So, based on the premise that dinosaurs are a fiction designed to disprove creationism and drive us away from God, the writers of this website go into extremely convoluted thinking about dinosaurs and paleontology. A long section uses quotes out of context from the Berkeley evolution website to claim that scientists dreamed up the whole thing as a big scam to undermine religion. Never mind the fact that all the early dinosaur discoveries were made by religious people such as Gideon Mantell, the Reverend William Buckland, Mary Anning, and Richard Owen and that many later paleontologists (like Edward D. Cope) were also quite religious. This writer knows how to clip little bits of simplistic web histories out of context but doesn't know enough history to know the difference.

The next section on the website is another long, bizarre example of quote mining, where the author clearly knows nothing whatsoever about fossils and the way they are found. The author jumps from one paranoid speculation to another, all in an attempt to suggest that dinosaur bones are forgeries planted in the outcrop by crooked paleontologists and there is no way they could have gotten there without fraud. The list of mistakes, lies, and misconceptions about fossils and geology is so long that I don't have space to even begin listing them all. Because fossil skeletons are incomplete in the field, it is common practice to mold replicas to complete the skeleton for display. But the author of this website then jumps to the absurd conclusion that *all* the bones in *every* dinosaur skeleton on display are faked!

From there, this site goes into the old shopworn and long-debunked creationist attacks on radiometric dating (see chap. 12) and geology, using the classic tactic of quoting out of context to show the opposite of what the text really intended. Then the author savages other creationists, including those who use the great size of dinosaurs to justify an expanding earth with stronger gravity today than in the past and others who quote the passages about the "Behemoth" in the Book of Job as evidence that the Bible talks about dinosaurs.

Then the site jumps into other debunked ideas, like Tom Gold's abiogenic origin of petroleum (long ago falsified), and presents a list of nineteenth-century naturalists who talked about dinosaurs and claims that they allegedly made up their discoveries in order to promote evolution! Once again, this guy is so ignorant of history that he has no idea that half the list consisted of devout individuals who were creationists and most of them worked on dinosaurs in a religious context. Not only that, but they were not trying to "prove evolution"; their work was done decades before Darwin's book came out in 1859. Give him an F in history.

Finally, he trots out the laughable idea that paleontologists concocted this whole forgery to get rich, but clearly he knows nothing about real paleontology. Most of my colleagues in paleontology have turned down the opportunity to get an education in more lucrative careers in law, medicine, or business to work on fossils at a mere fraction of the salary that they could be getting elsewhere. Nor do professional paleontologists get rich from selling their fossils; the only people who get rich from fossils are commercial collectors, who are not professionally trained or doing research but simply in it for the money.

So how can anyone become so misguided and get so many things wrong that are easy to check? It turns out that many creationist communities are highly insular and not only avoid secular media but only hear and read what their church leaders tell them to. If you read only creationist ideas for a long time and surf websites looking for things that you can quote out of context, you too can become someone who can manage to get every fact in the article 100 percent wrong! As long as your faith in biblical literalism is more important than checking the facts out for yourself, you can twist anything to suit this delusional worldview.

Read this website if you dare. You need a strong stomach for lies and self-deception, and hopefully you will not be shocked by the low view of humanity that emerges from reading it. And you will find that this is not the only place with such thinking. There are several others like it, such as Clinton B. Ames of www.swordofthevaliant.com ministries, who also claims on YouTube that dinosaurs are faked.[15]

What they write is so appalling and shows no actual firsthand experience with real fossils that it makes me despair for the future of humanity. But their ideas are no stranger than those of the modern-day geocentrists, flat-earthers, flood geologists, and many other extreme creationists who believe incredible nonsense just because the Bible says so. Heck, even my good friend Bill Nye got booed in Waco, Texas, by people who were shocked when he told them that the moon's light is merely reflected from the sun, since the Bible says God put "two great lights" in the heavens.[16] Only a few centuries ago, some scholars thought that fossils were works of the devil placed there to challenge our faith, and the widespread acceptance of the idea that they are the remains of once-living organisms is less than 250 years old. Our modern age of scientific enlightenment is a very young phenomenon in the scope of human thought and apparently a very fragile one in the face of fundamentalist religion.

Once again, we are seeing the bizarre looking-glass world of the internet and the people who haunt it, where black is white, up is down, faith is fact, evil is good, science is bad, and people will believe just about any wacky idea they encounter. We are also in the tricky position demonstrated by all Poes: the fringe believers out there are so bizarre that it's often impossible to tell a well-crafted parody from the real thing.

(12)

Is the Earth Only Six Thousand Years Old?

It's about Time

For centuries, different cultures have had different ideas about how old the earth is. Many, such as the Hindu and Buddhist belief systems, think of time as cyclical and continuous, with no beginning or end. Many other cultures date the beginning of the earth from a creation myth, but the amount of time since creation is usually not specified. The cultures influenced by the Abrahamic religions (Judaism, Islam, and Christianity) have a very distinct creation event (the seven days of creation mentioned in Genesis 1), although it is tricky to interpolate from the Bible how long ago that series of events occurred.

The idea that the earth is only six thousand to ten thousand years old is a relatively recent notion, first proposed by biblical scholars in the Middle Ages and later. By the 1600s, it was considered the pinnacle of scholarship to try to calculate the age of the earth using clues from the Bible. However, doing so is trickier than you might think. The Bible gives a chronology of some events, and there are ages of known historical events in Israel, Egypt, and Mesopotamia to help calibrate the chronology, but it's not enough. There are too many gaps and unrecorded intervals of time in the Scriptures between the "begat" verses, such as "and Methuselah lived after he begat Lamech seven hundred eighty and two years, and begat sons and daughters" (Genesis 5:26), to allow a precise calculation of the age of the earth from biblical texts. Nevertheless, using the begat method and correlating biblical events with what was then known of history in Egypt and Mesopotamia, scholars came up with estimates of the earth's age at around six thousand years old.

The most famous of these estimates was done by Archbishop James Ussher, an Anglican clergyman whose bishopric in Armagh, Ireland, was mostly Catholic, so there were very few Anglicans to minister to. His interest was mostly biblical scholarship, and his calculation of the age of the earth was the most influential of all. In 1647, he used all the information from ancient historical events to calibrate his chronology, but the oldest that could be dated back then was the reign of Babylonian king Nebuchadnezzar, about 600 BCE. He was also faced with the inconsistencies of the various versions of the Bible, so he had to choose a version that worked with what he thought made sense. Another consideration was the notion that Solomon's Temple was supposedly completed three thousand years after the creation. Jesus, the "fulfillment of the Temple," was supposedly born about one thousand years after Solomon's Temple, so that gave Ussher roughly four thousand years from before Jesus's birth. Ussher used all these calibration points as well as the fact that King Herod died in 4 BCE. This meant that Jesus had to be born four years earlier than his conventional birth year when the BCE/CE (or BC/AD) split, because his parents fled to Egypt to escape Herod's persecution after he was born. So if the Bible is to be taken literally, Christ was born four years "before Christ."

Putting all this together, Ussher used the four thousand years from creation to Solomon's Temple to Jesus's birth in 4 BCE and came up with a date of creation at October 22, 4004 BCE. Another scholar, John Lightfoot, amplified this date and claimed that creation occurred at 9:00 a.m. How anyone could decide what time of day it was when there was no sun in the sky yet or what date or month it was before there were calendars and seasons is an unanswered question. Nevertheless, these estimates were considered the paragon of good scholarship in the 1600s. Although they have been gradually abandoned by most Christians since then, they still appear in the marginal notes of some Bibles and are pushed by the Young-Earth Creationists (YECs). Some YECs do not strictly go with Ussher's estimate of six thousand years and concede that it might be as long as ten thousand years, but none of them will push the age of the earth much older than that. Because they take the Bible literally, they think it specifies ages in this range, even though it does not actually supply the age of the earth and only Ussher's calculations and questionable assumptions give any sort of date at all.

For much of the twentieth century, most of the Christians who believed that the creation week was a real event (and not a metaphor or myth) also subscribed to ideas that allow for a much older earth. For example, as discussed in chapter 10, Day-Age creationists consider each day in Genesis 1 not to be a literal twenty-four-hour day (since there was no sun to mark time with). Instead, each day is a long age that allows for the millions of years of geologic time. During the middle of the twentieth century, they were a major influence on the creationist movement. One of the most prominent of these was Dr. Lawrence Kulp, one of the experts on radiometric dating techniques who was also a devout Day-Age creationist. These creationists have since been pushed aside by the strident and activist YEC movement, so you seldom hear anything from them now.

"No Vestige of a Beginning"

In chapter 10, we discussed the ideas of James Hutton, the Scottish "Father of Modern Geology," and the way that they revolutionized geology. One of his biggest contributions was the recognition of the enormity of geologic time. Hutton saw the incredibly slow process of weathering and erosion at work, especially when he studied how the soils in his properties weathered and eroded. He examined Hadrian's Wall, built by the Romans around 128 CE across the entire English border with Scotland to protect Roman England from Scottish warrior raids. Hutton noticed that it had scarcely weathered in sixteen centuries. From this, he inferred that weathering of whole mountain ranges must take immense amounts of time. When he saw a stack of layered sedimentary rocks, he realized that they must represent not the instantaneous deposits of Noah's flood but instead the remnants of entire mountain ranges slowly eroded away or of seas that had slowly risen and flooded the continents.

His most important discovery was what geologists call an angular unconformity: a series of layered rocks which were steeply tilted or folded, covered by a surface that eroded them flat, and then covered by another nearly horizontal series of layered rock (fig. 12.1). Hutton reasoned that this could have only formed by a lot of very slow processes. First, the lower sequence of rocks was deposited by rivers or by seas as soft sands and muds; then it was hardened into rocks like sandstone and shale; then it was tilted by immeasurably great forces that uplifted mountain ranges; then the mountain range must have eroded down, forming the surface of erosion that geologists call an unconformity; and then the seas slowly rose again, drowning the ancient erosional surface and covering it with another flat-lying sequence of layered sandstones and mudstones. Because Hutton could see that each of these events took enormous amounts of time, the entire angular unconformity must represent millions of years' worth of time, not the six thousand years

Figure 12.1. Hutton's famous angular unconformity at Siccar Point, Berwickshire, on the southeast coast of Scotland. The lower sequence of rock layers has been tilted nearly vertically, then eroded off after having been uplifted into a mountain range. When the mountains finally wore down, the erosional surface was then covered by deposits of younger layered sedimentary rocks at an angle to the old sequence. Hutton correctly realized that such an outcrop required enormous amounts of time to form. (*Photo by the author.*)

granted by some biblical scholars. In Hutton's words, there was "no vestige of a beginning, no prospect of an end." The earth was not a few thousand years old but so old that there was "no vestige of a beginning." With this discovery, the concept of geologic time, or deep time, in millions and billions of years, was born.

The Dating Game

James Hutton wrote that the earth had "no vestige of a beginning," and almost all geologists since Charles Lyell in 1830 have agreed that the earth is immensely old. But just how old? How could we put a number on the age of the earth?

The problem was a challenging one, but it didn't daunt scientists, who tried all sorts of ingenious solutions. The most common method was to add up the total thickness of all the sedimentary rocks on earth, calculate how long it would have taken to deposit them, and use that as a minimum age. Most such estimates came to about one hundred million years since the Cambrian, which we now know is off by a factor of almost fifty. Why? As in all these early methods, a faulty assumption was built into the calculation. The most important problem is that they did not account for erosional gaps in the record, or unconformities, when there is no rock representing a time interval. Later studies have shown that the rock record is full of many time gaps and is actually "more gaps than record." Some geologists at the time suspected that there might be a problem with unconformities, but back in the 1800s, no one could know how big a problem it really was.

Then there was the famous estimate by Irish physicist John Joly. He tried to calculate how long it would take for the original oceans to go from fresh water to their current salinity, knowing the rate at which salt enters the oceans from the world's rivers. He also came to estimates of eighty to one hundred million years, which we now know to be off by more than a factor of fifty. What went wrong? The problem was his faulty assumption that the oceans have been constantly adding salt since they formed. It turns out that the salt content of the oceans does not change much through time, because much of it gets locked into salt deposits in the earth's crust and seawater salinity is in equilibrium and stays very stable.

But the most infamous and influential estimate was made by the famous physicist William Thomson (later known by his title, Lord Kelvin). Kelvin made huge discoveries in the fields of physics, especially thermodynamics. The Kelvin temperature scale is named after him, since he pioneered the concept of absolute zero temperature (now known as 0 K). He was also a great inventor and helped create the transatlantic cable system that established telegraph and then telephone communications between Europe and North America. He was a giant among scientists of his time, and few people dared disagree with him.

In 1862, Kelvin attacked the problem of the age of the earth using thermodynamics. He assumed that the earth had started as a molten ball at the same temperature as the sun and that it had cooled off at rates that we can measure from the heat coming up from the earth's interior. From this method, he estimated the earth was only twenty million years old, much younger than most geologists were willing to accept. This was also a problem for Charles Darwin, who knew that the earth had to be immensely old for his newly proposed concept of evolution to work, as Kelvin's estimate didn't seem to offer enough time.

Through the rest of the nineteenth century, physicists and geologists were at an impasse. Neither could comprehend the arguments of the other side or see the flaws in their own estimates. By the late 1800s, geologists began to give in and fudge their estimates a bit from the original values of eighty to one hundred million years to numbers closer to Kelvin's twenty million years. Physics envy was just as powerful then as it is now! But the problem with Kelvin's estimate was just like the others: faulty assumptions. Kelvin assumed that the heat was from the original solar system and that no other heat sources were involved. We now know this is wrong. The earth does have an additional heat source.

In 1896, Henri Becquerel in France discovered radioactivity, and in 1903, Marie and Pierre Curie showed that radioactive materials like radium produced a lot of heat. At that time, New Zealander Ernst Rutherford was England's foremost authority on this new source of energy. In 1904, he was getting ready to address the Royal Institution of Great Britain about this new discovery when he suddenly realized that the eighty-year-old Lord Kelvin himself was in the audience! The young Rutherford was about to challenge the world's most famous physicist's estimate of the age of the earth! As Rutherford wrote later,

> I came into the room which was half-dark and presently spotted Lord Kelvin in the audience, and realised that I was in for trouble at the last part of my speech dealing with the age of the Earth, where my views conflicted with his. . . . To my relief, Kelvin fell fast asleep, but as I came to the important point, I saw the old bird sit up, open an eye and cock a baleful glance at me. Then a sudden inspiration came, and I said Lord Kelvin had limited the age of the Earth, provided no new source [of heat] was discovered. That prophetic utterance referred to what we are now considering tonight, radium! Behold! The old boy beamed upon me.[1]

Kelvin's estimate had been based on the faulty assumption that there were no other sources of heat beyond the earth's original heat when it cooled from a molten mass and that it would cool in no more than twenty million years. But radioactivity provides that additional heat. In fact, radioactivity provides so much heat that it is now the only source of heat that we measure coming from the earth's interior. The original heat from the cooling of the earth Kelvin thought he was measuring dissipated billions of years ago, maybe even during the first twenty million years since the earth first formed 4.6 billion years ago.

Becquerel, the Curies, and Rutherford had pioneered the physics and chemistry of radioactivity, and their discoveries showed that Kelvin's assumption of no additional heat was wrong. But they were not geologists interested in dating the earth. It was two other scientists, Bertram Boltwood and Arthur Holmes, who realized that radioactivity provided a solution not only to Kelvin's dilemma of where the heat came from but also the answer to the question, how old is the earth?

The method is a simple one but widely misunderstood. There are just a handful of elements in nature that are radioactive, and they spontaneously decay from a parent atom (such as uranium-238, uranium-235, and potassium-40) to a corresponding stable daughter atom (lead-207, lead-206, and argon-40, respectively) at rates slow enough to use in geologic dating. This rate of decay is precisely known, so if we can measure the amount of both the parent atom and the daughter atom in a sample, their ratio is a measure of how long that decay has been ticking away.

Of course, there are lots of complications with real rocks, so very special conditions have to be met. As the decay is measured from the time the decaying parent atoms are first locked into a crystal, it primarily works on igneous rocks that cool from a magma, like lava flows, volcanic ash layers, and intruding bodies of magma forming dikes. Geochronologists (specialists in radiometric dating) try to get the freshest crystals possible so that there is no leakage and contamination of parent or daughter atoms leaving or entering the crystal that might distort the ratio. There are all sorts of laboratory procedures designed to eliminate expected problems in advance and to correctly calibrate their instrument (known as a mass spectrometer, since it separates and measures the different isotopes by mass) to give reliable ages. Finally, every radiometric date comes with an error estimate, based on how reproducible each result from the machine can be. So, for example, if they cite an age of 100 m.y. ago ± 5 million years, they are saying that with a 95 percent probability, the true age lies somewhere between 95 m.y. ago and 105 m.y. ago.

But none of this was known in 1900 when radioactivity was just beginning to be understood. Physicists had been trying to date rocks by measuring the helium given off by uranium decay, but it was almost impossible to capture all the helium gas. Instead, it was Yale chemist Bertram Boltwood who discovered that uranium decayed to lead through radioactive breakdown. Following on a suggestion from Rutherford, Boltwood noticed that the rocks he knew to be older had more lead in them than those that he knew to be younger. Unfortunately, he was using the very primitive understanding of the uranium-lead (U-Pb) system that prevailed at the time. He didn't realize, for example, that there are two different radioactive isotopes of uranium, uranium-238 and uranium-235, each with a different decay rate and a different daughter isotope of lead. Nonetheless, he analyzed the samples he had, and in 1907, he got samples ranging from 400 million years to as old as 2.2 billion years. This was the first evidence that the earth was indeed billions of years old, as geologists had long suspected, and that Kelvin's estimate was way off. Unfortunately, in his later life, Boltwood suffered from severe depression, and his research came to a standstill. He died by suicide in 1927, unable to cope with his personal problems.

Boltwood had done the first analyses and gotten ages on rocks as old as 2.2 billion years, but he never followed up on his breakthrough. Thus, it fell to a younger British geologist, Arthur Holmes, to take the budding young field of geochronology and develop it into a rigorous science. Holmes latched on to the hot problem of radioactivity and realized that Boltwood's 1907 paper on uranium-lead dating held immense promise. For his undergraduate research project, he had a granitic rock from the Devonian of Norway to analyze. He cut his Christmas holiday short and spent his holiday break in the cold, dark lab in London, working all by himself in the silence. As his advisor, physicist Robert Strutt, later recalled, "We are at present largely subsisting on loaned apparatus, some of which belongs to other public bodies, such as the Royal Observatory, the Royal Society, etc., while some has been borrowed from private friends. I need hardly say that it seems rather below the dignity of an institution like the Imperial College that its teachers should have to beg apparatus of their personal friends for the purpose of teaching the students."[2]

As Lewis described it,[3] Holmes worked away in the cold, silent, lonely lab in January 1910, crushing the rock into a powder in a mortar made of agate, fusing the mineral grains in a platinum crucible with borax, dissolving it in extremely caustic hydrofluoric acid, and then boiling it again and again while measuring the radon emissions (an indirect measure of how much uranium was present). The lead content was measured by fusing the powder in a cake, boiling it, twice dissolving it in hydrochloric acid, and letting it evaporate all its water. Then he heated it in ammonium sulfide to make the lead precipitate out as lead sulfide (known as the mineral galena). The precipitate was collected on a filter, dried, ignited, treated with nitric acid, boiled, treated with sulfuric acid, and heated again. Eventually, as Holmes wrote, "A tiny white precipitate then remained. This was collected on a very small filter, washed with alcohol, dried, ignited, and weighed with the greatest possible accuracy."[4] Often there were only a few milligrams of material left.

These complicated chemical operations took incredible patience, extraordinary dexterity, and lots of time, and they used up nearly all his original sample. Then to top it off, the results had to be verified, so the entire analysis was repeated two to five times, depending on how much original sample he had left to work with. Once Holmes had to discard all his data because radon had been leaking into the room. In other cases, he had to go begging to the British Museum for more samples, because he had used up his original allotment. Eventually, however, all this hard work paid off, and Holmes got a reliable date of 370 million years old on his Devonian granite from Norway. He had greatly improved Boltwood's original methods and proven that uranium-lead dating could work and that it was possible to generate dates on rocks. The results were published in 1911, soon after Holmes graduated in 1910.

By 1913, Holmes had so many new results and so many improvements on the method that he was able to write his groundbreaking book, *The Age of the Earth*, while he was still a graduate student. In it, he not only explained the basic principles of geochronology and discussed the problems with earlier methods of dating the earth; he also finally laid Lord Kelvin's mistaken estimate to rest. He had dates on some of the oldest rocks in Britain of 1.6 billion years old, although he refused to speculate on the age of the earth. Later editions included results from analyses of older and older samples, until by the 1950s, he had dates of 4.5 billion years on different meteorites, our present estimate of the age of the earth.

Holmes's early research earned him his doctorate in 1917 from University College London. Eventually, after several years trying to make a living as an exploration geologist in Africa and Asia (and losing his infant son to disease), his reputation landed him a teaching and research job at Durham University in northern England, close to his birthplace. There Holmes spent the next eighteen years teaching geology and adding to and refining the database of radiometric dates from around the world before finishing his career at the University of Edinburgh. His work was so dominant in the field that he later became known as the "Father of Geochronology" or the "Father of the Geologic Time Scale." Before he died in 1965, Holmes received the Vetlesen Prize, the "Nobel Prize" of geology, for his work.

Impervious to Evidence

The YECs, of course, cannot admit that the earth is more than ten thousand years old, so they resort to any method possible to try to discredit the evidence for the age of the earth. Most of the time, they find a tiny inconsistency in the dating or criticize any assumption that they can in order to suggest that no dating method in geology is reliable. From this, they think that we are forced to accept their conclusion that the earth is less than ten thousand years old.

Of course, it doesn't work that way. Like any scientific field, there are good data and bad data, and scientists try their best to eliminate the problems and indicate the uncertainties of dates. Creationists don't give scientists credit for being skeptical and self-critical about their own data. But anyone who deals with geochronology knows that the dates are subject to constant scrutiny by multiple labs and that anything fishy is quickly challenged and rejected. The result is an extremely robust set of data. Multiple independent radioactive atomic systems (for example, potassium-argon [K-Ar], uranium-lead, and rubidium-strontium [Rb-Sr]) are used on the same samples, so if any one of them is giving problems, it clearly can be thrown out. Creationists point to one or two examples of supposedly unreliable dates, but when three or more independent dating methods are run in different competing labs on the same rock and give the same answer, there is no chance that this is an accident! After nearly a century of analyses, with thousands of dates checked and rechecked like this, geologists are as confident about the reliability of radiometric dating as they are about any other field of science. Creationists will mention a specific date that proved to be wrong as evidence that the entire field of geochronology is unreliable, when in fact it was the geologists themselves who spotted the erroneous date and quickly rejected it.

As my friend Brent Dalrymple (the leading expert on potassium-argon dating) points out, it's as if we were in a clock shop. There is a wide variety of watches and clocks in the shop, a few of which don't keep accurate time. But that fact doesn't mean we completely ignore clocks and watches altogether, as creationists are doing by rejecting *all* radiometric dating out of hand. We simply keep checking them against each other to determine which ones are reliable and which ones are not. Creationists view the entire method of radiometric dating as suspect if there is but a single bad date; this is an extremely Manichean, black-and-white, all-or-nothing position. But in science (as in anything in the real world), the situation is not black and white but shades of gray. There are things we have 99 percent confidence in (like the fact that you will almost certainly die if you jump off a high building) and things that we have less confidence in. Geochronologists publish the error estimates of their dates, so we know how much confidence they (and we) should have in any particular age estimate. As the old saying goes, "One bad apple don't spoil the whole barrel." One bad date does not discredit the entire field of geochronology.

How Do We Know?

As we have in other chapters, let us look at specific claims that YECs use to suggest a young earth and then discuss why their ideas are not taken seriously by real scientists.

1. In one commonly repeated claim, creationists mock geologists about a living clam that gave a radiocarbon date of many thousands of years old. But this is a well-understood anomaly. Normal radiocarbon dating works when nitrogen-14 from the atmosphere is transformed by cosmic radiation into carbon-14, which is then incorporated into living tissues. When the organism dies, the carbon-14 begins to decay, making the bone or shell or wood (or anything bearing carbon) datable, as long as it is less than eighty thousand years old, since the radiocarbon decay rate is relatively fast. These peculiar clams live in water covering ancient limestone that releases radioactively dead carbon into the water. That ancient carbon is then incorporated into the mollusk shells, where it

throws the ratio off. Radiocarbon specialists have long been aware of this minor problem, and they never rely on dates where this kind of contamination could be an issue.

2. A creationist named Thomas Barnes argued that the earth's magnetic field is decaying and that extrapolating the field strength back in time suggests that the earth is less than ten thousand years old. But as we saw in chapter 7, the long-term behavior of the earth's magnetic field is very well known. It fluctuates in strength and direction over thousands of years, which we can determine by measuring the magnetic intensity recorded in ancient rocks. Barnes assumed a simple linear change through time, but he ignored the abundant scientific evidence of the increases as well as decreases in the earth's field strength.

3. Henry Morris attempted to calculate the age of the earth by estimating how long it would take for 3.5 billion people (now over 7.8 billion) to arise from Adam and Eve. From present-day rates of population growth, it seems like all humans date back only a few thousand years. But this completely ignores the fact that human populations, especially in the distant past, didn't increase exponentially like bacteria in a Petri dish. There is very solid archeological evidence that for most of human history, human populations remained roughly constant for thousands of years due to the restraints of death, warfare, and disease. Only in the past fifty years has the population explosion come to resemble that of bacteria. Morris's extrapolation is entirely unsupported by the data.

4. Morris is fond of citing figures that indicate that about five million metric tons of cosmic dust fall on the earth each year. He calculates that over five billion years, we should have accumulated a layer of dust 55 meters (182 feet) thick! But if you do the calculation correctly (which Morris does not), it amounts to only a shoebox full of dust over an entire square kilometer. This is so minuscule that it can barely be detected even in the best sedimentary records from deep-sea cores, which are undisturbed and have extremely low sedimentation rates and are the only places so little dust could be detected. In shallow marine or terrestrial sediments, which are highly mixed and weathered, this tiny amount of dust would be homogenized quickly.

5. Creationists such as Ken Ham are fond of claiming that there is a site in Australia with fossilized trees that have been potassium-argon

dated by volcanic ash around 45 million years old, but when they are dated with radiocarbon, they are only 40,000 years old. For them, this is proof that *no* radiometric dating method can be trusted. The thing to remember is that each radioactive clock ticks at a different rate, but they all keep good time—*if* they are used properly. Radiocarbon has a very short half-life of 5,370 years, so it decays extremely quickly. By 40,000 years (some labs can push it up to 60,000–80,000 years now), an object is radiocarbon dead; no more decay is occurring, and you cannot use it to measure *anything* anymore. Thus, radiocarbon is primarily used by scientists who work on really young events of the last Ice Age: archeologists and geologists who work on the last glacial cycle, or the last 10,000 years since the last Ice Age ended.

A real scientist would never even *consider* using it for anything older, and any fool who does so shows that they have no clue what they're doing. Other isotopic systems are useful in different age ranges. U-Pb (both isotopic pairs) and Rb-Sr decay over billions of years, so they are only used on the oldest earth rocks, plus moon rocks and meteorites. K-Ar is the system used by most geologists, as it can date rocks from less than one million years old to the oldest rocks we have. Thus, its useful age range covers the vast majority of common geologic settings. To go back to our clock analogy, radiocarbon is like a clock that ticks really fast and runs down quickly. U-Pb and Rb-Sr are like a big grandfather clock that ticks very slowly but doesn't run down except over a very long time.

In this case, their example of the Australian trees in lava only demonstrates the complete incompetence of creationists. They used K-Ar to date the lava at forty-five million years, whereas no real geologists would waste their time dating the radiocarbon-dead trees. But the creationists ran radiocarbon on them anyway and, sure enough, got dates of around forty thousand years, which only means the sample is radiocarbon dead and older than forty thousand years, not that the age of the sample is forty thousand years. It's like looking at the time on a fast-running clock that has stopped and comparing it to the slow grandfather clock. Creationists say that since they are different, *no* clocks can be trusted, when in fact they were foolishly looking at a clock that has run down and stopped.

6. There are living trees, such as the bristlecone pines of the White Mountains of California, that are over 5,000 years old, and a number of dead trees give us a tree-ring record that spans over 11,000 years. These trees are useful in giving a detailed climate record of the past 11,000 years—a real problem if you believe the earth is less than 6,000 years old. There is a colony of quaking aspen trees in the Fishlake National Forest of Utah that is over 80,000 years old, if you patch together the tree-ring records of the youngest living trees, to the oldest trees, and all the trees in between. There are clonal colonies of eucalyptus in Australia that are older than 13,000 years. Old Tjikko, a Norway spruce in Sweden, is dated over 9,550 years old. These and many other examples make it pretty hard to claim that the earth is only 6,000 years old.

7. Numerous ice cores have been recovered from Antarctica and Greenland that have bands of light and dark ice caused by the seasonal melting and freezing each winter and summer. We can verify this by looking at the top layers of the core and finding each dark and light band formed during the last few winters and summers. You can then count the bands on individual ice cores and determine how many years they go back into the past. The North Greenland Ice Project produced a core that went back over 140,000 years. Vostok Station on Antarctica, run since 1957 by the Soviet Union and now by Russia, has retrieved longer and longer cores, originally going back about 130,000 years, and their latest core goes back 436,000 years. These cores not only record hundreds of thousands of years of snowfall but also trap individual bubbles of air from the distant past that allow us to measure their gases like carbon dioxide and determine the temperature for every year in the past. That record was beaten in 2004, when the European Project for Ice Coring in Antarctica (EPICA) recovered a core from Dome C near Dronning Maud Land in East Antarctica. Known as EPICA-1, it goes almost 800,000 years into the past, or eight complete glacial-interglacial cycles of the ice ages. It's pretty hard to imagine how this can be squeezed into a time scale of only 6,000 years unless a creationist is claiming that there were about 133 winter-summer cycles per year (or one every two to three days) and eight complete glacial-interglacial cycles in only 6,000 years. Antarctica doesn't even have that many snowstorms in a decade, let alone that many winter-summer seasonal cycles!

There are many other examples of geologic systems that give reliable dates older than six thousand by a wide variety of measurements, but it is pointless to belabor the point any further. Suffice it to say that each creationist attempt to discredit the estimates of the age of the earth boils down to special pleading and ad hoc hypotheses, shifting the goal posts, lying and misrepresenting the data (sometimes accidentally but often deliberately), and, in most cases, completely ignoring any data they can't explain. No real scientist would be allowed to get away with such unethical and dishonest behavior.

(13)

Mysteries of Mount Shasta

Mighty Mount Shasta

Mount Shasta is one of the most mystical and awe-inspiring sights in California, if not the world. It is California's biggest volcano (fig. 13.1) and rises to 4,322 meters (14,179 feet) in elevation. Second only to Mount Rainier among Cascade peaks, it is the fifth-highest mountain in California after four Sierra peaks. In many ways, it is more impressive than any other peak, since it is built of 354 cubic kilometers (85 cubic miles) of rock, bigger in volume than Mount Rainier or any other Cascade volcano. Mount Shasta towers 3,000 meters (10,000 feet) above the surrounding landscape with no other volcanoes or other Sierra-like peaks nearby. On a clear day, its snow-capped summit can be seen for 230 kilometers (140 miles) to the south down in the Sacramento Valley.

Many people have been impressed and mystified by it. The pioneering naturalist John Muir wrote, "When I first caught sight of it over the braided folds of the Sacramento Valley, I was fifty miles away and afoot, alone and weary. Yet all my blood turned to wine, and I have not been weary since."[1] The Shasta tribe talks about it in their flood legends: "At last the water went down. . . . Then the animal people came down from the top of Mount Shasta and made new homes for themselves. They scattered everywhere and became the ancestors of all the animal peoples of the earth."[2] And in 1908, President Theodore Roosevelt wrote, "I consider the evening twilight on Mount Shasta one of the grandest sights I have ever witnessed."[3]

Such an impressive edifice naturally inspires myths and legends. The Klamath peoples who lived around it thought the mountain was the home of the Spirit of the Above-World named Skell, who had come down from heaven to live on top of the mountain at the request of the ancient chiefs of the Klamaths. Skell fought with Llao, the spirit of the Below-World, who lived on Mount Mazama (since erupted and exploded to form Crater Lake). Skell

Figure 13.1. Photo of Mount Shasta. The main summit and the smaller peak Shastina give it a saddle-shaped profile. A young volcanic cinder cone is visible in the foreground. (*Courtesy Wikimedia Commons.*)

and Llao threw rocks and lava at each other, which the Klamath peoples had seen thrown out of the mountain during ancient eruptions.

In more recent years, Shasta has been the focus of many different spiritual and paranormal believers. Modern descendants of the Klamath tribe and other Native American groups routinely meet in the McCloud River area to worship and practice their rituals. The Buddhist leader Houn Jiyu-Kennett established a monastery in 1971 called Shasta Abbey, where a large group of Buddhist monks and many lay visitors observe a strict schedule of meditation, silence, and serenity, seeking to find the Middle Way.

In addition to these traditional forms of worship, the mountain has become a center for a lot of more unconventional New Age beliefs, from Bigfoot claims, to claims of alien abductions and UFOs landing on and even in the mountain, to even wilder ideas. One of the strangest is the Lost Continent of Lemuria.

Lemuria

The legend of the Lost Continent of Lemuria actually originated as a legitimate scientific hypothesis, but then it morphed into New Age mythology. The idea was first proposed in 1864 by the pioneering zoologist Philip Sclater to explain the presence of primitive primate fossils related to lemurs in both Madagascar and India but not in Africa or the rest of Eurasia. He suggested that there was once a lost continent of Lemuria that formed a land bridge across the Indian Ocean from Madagascar to India, which has since sunk into the ocean. In Sclater's words, "The anomalies of the mammal fauna of Madagascar can best be explained by supposing that . . . a large continent occupied parts of the Atlantic and Indian Oceans . . . that this continent was broken up into islands, of which some have become amalgamated with . . . Africa, some . . . with what is now Asia; and that in Madagascar and the Mascarene Islands we have existing relics of this great continent, for which . . . I should propose the name Lemuria!"[4]

After its proposal, Lemuria became a popular concept among scientists of the late 1800s, especially as a device for evolutionary explanations. The great German biologist Ernst Haeckel, Darwin's chief defender in Germany and a pioneer of both micropaleontology and embryology, blamed the sinking of Lemuria for the absence of "missing link" fossils that connect the lineages of primates together with the origins of humans. Other scientists transferred Lemuria to the Pacific Ocean, again as a possible land bridge to explain the transpacific distribution of some organisms, even though lemurs were no longer involved.

Sclater's proposal wasn't outrageous for its time, when many other biologists were proposing sunken land bridges as a means of explaining the distribution of closely related organisms on either side of an ocean. But the entire idea became obsolete when plate tectonics came along in the 1960s and 1970s, and the floor of the Indian Ocean was finally mapped in detail. No such "sunken continent" exists on the Indian Ocean floor; the only large raised features are the uplifts around the island of Mauritius, the Kerguelen Plateau in the south-central Indian Ocean, and the plateau around New Zealand called Zealandia, but these don't form any kind of land bridge from Madagascar to India. In fact, there was once a land connection between India and Madagascar, but this was when both were part of the supercontinent of Gondwana, from the Early Permian (about three hundred million years ago) until the Late Cretaceous (about eighty million years ago). There are many similarities among the Cretaceous Gondwana dinosaurs, including those found not only in Madagascar and India but also in Africa, South America, and sometimes even in Antarctica and Australia, all remnants of Gondwana. But by the early Eocene (fifty-five million years ago), when some of the primitive lemur fossils that Sclater was studying lived, India had long since broken away from Madagascar and the rest of Gondwana, so its past Gondwana connection is no help in explaining the distribution of lemurs and their fossil relatives, the primary reason for Sclater's proposal in the first place.

After zoologists abandoned the idea, Lemuria no longer had any scientific meaning and became closely associated with New Age philosophy, the cult of theosophy, and many other fringe beliefs. Lemuria has become a popular concept in science fiction and fantasy literature as well. In this case, Lemuria no longer has any real connection to an actual place on earth but serves as one of the universal missing continents/civilizations that are so common in occult and paranormal literature.

Not surprisingly, Mount Shasta and Lemuria soon appeared together. According to one site,

> In the 1880s a Siskiyou County, California, resident named Frederick Spencer Oliver wrote *A Dweller on Two Plants, or, the Dividing of the Way* which described a secret city inside of Mount Shasta, and in passing mentioned Lemuria. Edgar Lucian Larkin, a writer and astronomer, wrote in 1913 an article in which he reviewed the Oliver book. In 1925 a writer by the name of Selvius wrote *Descendants of Lemuria: A Description of an Ancient Cult in America*, which was published in the *Mystic Triangle* [a Rosicrucian publication], Aug., 1925 and which was entirely about the mystic Lemurian village at Mount Shasta. Selvius reported that Larkin had seen the Lemurian village through a telescope. In 1931 Wisar Spenle Cerve [a pseudonym for Harvey Spencer Lewis] published a widely read book entitled *Lemuria: The Lost Continent of the Pacific* in which the Selvius material appeared in a slightly elaborated fashion. . . .

According to Zanger, Frederick Spencer Oliver was a Yrekan teen who claimed that his hand began to uncontrollably write a manuscript dictated to him by Phylos, a Lemurian spirit. Meisse points out that Oliver's novel of spiritual fiction is "The single most important source of Mount Shasta's esoteric legends. The book contains the first published references linking Mt. Shasta to: (1) a mystical brotherhood; (2) a tunnel entrance to a secret city inside Mount Shasta; (3) Lemuria; (4) the concept of "I AM"; (5) "channeling" of ethereal spirits; (6) a panther surprise. The author claims to have written most of the novel within sight of Mount Shasta, and autobiographical telling of the story from Phylos the Thibetan's point of view is an interesting twist. . . .

In 1908, Adelia H. Taffinder wrote an article, "A Fragment of the Ancient Continent of Lemuria," for the *Atlantic Monthly*. In her article she links the concept of Lemuria to California, and Meisse proposes that the article, "with its Theosophical teachings and extension of the Lemurian Myth to California, may have been part of the research material involved in the creation of the Mount Shasta Lemurian Myth as presented by Selvius in 1925 and Creve in 1931."[5]

Much of the mythology of the theosophy cult and Mount Shasta has since passed on to other fringe religious cults, including the Summit Lighthouse and the Church Universal and Triumphant (a UFO cult founded by Elizabeth Clare Prophet), and Kryon (an "entity from beyond" who allegedly speaks to channeler Lee Carroll). I AM Activity (a cult offshoot of theosophy founded by Guy Ballard in Chicago in the 1930s) supposedly started when the founder was hiking on Mount Shasta and allegedly met a man calling himself the Count of St. Germain who taught Ballard the basic beliefs of the I AM cult.

Another legend is that in 1904, a British prospector named J. C. Brown discovered a lost underground city beneath Mount Shasta.[6] The cave sloped downward for 11 miles, and the underground village was filled with gold, shields, and mummies, some over 3 meters (10 feet) tall. When he told his story to John C. Root, they gathered an expedition in Stockton, but on the day they were to set out, Brown did not show up and was never heard from again.

In August 1987, a large gathering of over five thousand believers in the Harmonic Convergence came to Mount Shasta, convinced that something mysterious would happen, since it was one of the global "power centers."[7] They were inspired by a book by Jose Arguelles called *The Mayan Factor*, published in early 1987. He wrote that the prophecies of the Bible and the Mayan and the Aztec calendars predicted that on August 16–17, 1987, the world would come to an end or would start a new age. Arguelles argued that if four thousand self-chosen people were "resonating" with the harmony of the universe at that time, thinking peaceful thoughts, the world would be saved from destruction.

The region around Mount Shasta was crowded with thousands of believers, and campers and vehicles and tents were everywhere, including a lot of astrologers, channelers, and even some Bible-thumpers trying to convert them from their New Age beliefs. The fateful days came and went, and nothing was different except for the crowd of people around the mountain, who eventually all left. Whether they actually saved the world with their thoughts and prayers or not, the world went right on after the Harmonic Convergence.

UFOs on Shasta?

But the biggest paranormal belief about Mount Shasta centers on the claims that UFOs have landed there, that UFOs emerged from tunnels in Mount Shasta, and that aliens have been sighted there. Entire books have been written about these alleged sightings, and nearly every year now there is a conference about UFOs held in the nearby town of McCloud. A typical account of one of the ten speakers at the July 27–29, 2018, meeting (with about two hundred attendees) recounts the following:

Robert Potter, coordinator for the "From Venus With Love" conference, was dressed all in white as he spoke about how his life's journey started with an interest in pyramids in high school and, later, his contacts with space aliens and their extra-terrestrial teachings.

He and others spoke about their "time together for the purpose of true spiriting of brotherhood; the same message as Christ."

Potter said, "Until we stop wars and negativity, they (ETs) are staying at a distance. Civilization rises and falls with human destruction. They are willing to work with us to heal our planet. If they use their technology to clean up our pollution we would just pollute it again. They are guiding us towards abundance and prosperity and free energy. So many things would change if we welcome and embrace their offer of help."

He said, "Small groups and individuals, such as our speakers, are reaching out to be contacted. It is very positive, no fear. With so much information on the internet, people understand the movement and where it is going. With communication, education and following God's laws, we can live a long time. We believe in the science of being: Communication to connect with God. A healthy and balanced life with sleep, meditation and mental development. Financial service, to contribute to society. Socializing, communicate with others and not living alone."

Potter said his first contact was in 1952 with Orthon, Lady Orda and how Venusians look like us physically so not to be intrusive. They look to be about 22 to 29 years old but they are in actuality over 200 years."[8]

The internet is full of videos claiming to show UFOs (actually, just a series of lights in the night sky, with no critical analysis),[9] the "hangar door on the mountain" where UFOs enter and leave (actually, a deep shadow on a nearly vertical glacially carved cirque near the summit),[10] the "doorway to another dimension" (just a random hole in the rock),[11] and the perennial legends about secret underground cities beneath Mount Shasta.[12]

Most of these videos and accounts are the usual stuff by the UFO and alien believers: sketchy eyewitness accounts, blurry videos, credence given to old legends and stories of the paranormal with no skeptical scrutiny, and so on. (See my book with Timothy Callahan, *UFOs, Chemtrails, and Aliens: What Science Says* for further details). Given that so many other cults and New Agers believe that Shasta is a mystical mountain, it is not surprising that it is also a center for UFO fanatics. (Many of the New Agers are also in the UFO camp, so the groups are overlapping.)

But Mount Shasta has another phenomenon that leads people to believe they have seen UFOs over the mountain: the clouds above it. The clouds most commonly seen as "UFOs" in the sites linked above are known as lenticular clouds, or altocumulus clouds (fig. 13.2). Just type in "UFO" and "cloud" into your browser, and click on the "images" link and you'll see dozens of them, some of which are quite spectacular and do somewhat look like the disklike shape of a UFO. They get their name because they have a shape like a lens, a lozenge, a disk, or—yes—a flying saucer. Contrary to the rants of the UFO believers, these clouds are a well-understood meteorological phenomenon. They are the effect of fast-moving, upper-level horizontal air currents that flow upward from the lower level where the air is above dew point to a colder upper layer of air, where the air current cools and quickly

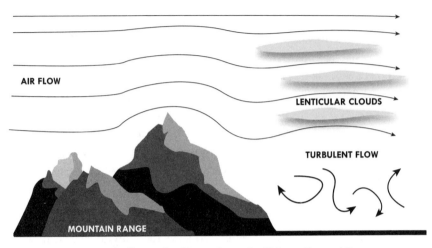

Figure 13.2. A lenticular "UFO cloud" near the peak of Mount Shasta. (*Courtesy Wikimedia Commons.*)

Figure 13.3. Lenticular clouds form when an obstacle such as a mountain diverts air currents upward until they cool below the dew point, allowing moisture to condense and form clouds. As the air currents sweep back down, they cross the dew point again and warm up, so the cloud is pinched off and ends up being lens shaped. If the perturbation makes a series of up and down air currents, a series of lenticular clouds may develop downwind of the disturbance. (*Courtesy Wikimedia Commons.*)

condenses (fig. 13.3). Then these same currents flow back down until they are above the dew point, leaving a little blob or hump of cloud behind where they were in the colder air.

Lenticular clouds are particularly common over the tops of high mountains, which divert air currents up into their colder tops, or just behind mountain ranges, which have created a wave-like pattern in the upper-level air currents that flow over them. Each time the wave of air rises into colder parts of the atmosphere, lenticular clouds are born (fig. 13.3). Some of the most spectacular and UFO-like lenticular clouds build over the peaks of the highest isolated mountains, especially tall conical volcanoes like Mount Rainier, Mount Shasta, and Mount Fuji (fig. 13.2). They form so often over Mount Shasta that the UFO believers are convinced that Mount Shasta is not a dormant volcano (which erupted as recently as two hundred years ago) but instead a secret base for alien operations.

Just type "Shasta" and "aliens" in to your browser, and you'll find another array of websites, all convinced that the aliens are hiding in Shasta and biding their time. (They never explain how the aliens hiding in the mountain survived its many eruptions or how they hollowed out an entire mountain made of hard rock and hot molten magma.) Other groups think that Mount Shasta is the site of a "harmonic convergence" where aliens and/or angels come to visit the earth and communicate with us.[13] This quote is typical:

> Now to give you a little more first hand knowledge. Lets take a look at some of the UFO lenticular clouds that are seen frequently over Mount Shasta which is located in California. Mount Shasta is a stratovolcano, also known as a composite volcano. There is an Spiritual Retreat on the Etheric Planes over the area between Sacramento and Mount Shasta, California. You can go there for guidance and to learn how to attain greater purity, discipline, and order. Students also learn strategies of light and darkness and how to connect with ascended beings by raising our human consciousness to the level of the Christ. This is not a physical place you go to. It is a place you go to while you are traveling out of body in Spirit. This spiritual retreat is run by Archangel Gabriel and Hope. You must also understand that Archangels, Ascended Masters and Celestial Gods are all extraterrestrial beings. This is why there are so many of the lenticular UFO clouds around Mount Shasta.[14]

What Shasta Is Really Like

We have seen a full range of New Agers, cultists, and UFO believers make up their own ideas about Mount Shasta. What none of them bother to consider is the geology of the mountain. The mountain is a volcano that erupted only two hundred years ago, and there is no evidence of a hollow mountain or tunnels within the mountain that the paranormal crowd believe in. In fact, the mountain has been climbed and hiked hundreds of times by many geologists, and all they ever see is the evidence of a variety of different volcanic rocks formed by many different events. No tunnels, no hangar door openings, nothing. In addition, the US Geological Survey monitors the mountain continuously,[15] looking for earthquakes that might indicate that magma is moving up into the mountain. It has been subjected to several gravity surveys, which would have clearly revealed tunnels, hollow chambers, or anything other than the solid mountain of volcanic rock that it really is.

Mount Shasta is not a simple conical volcano; it is built of four overlapping cones that combined to form a complex shape, including the main summit and a secondary peak called Shastina, which is 3,760 meters (12,330 feet) in elevation (fig. 13.1). Shastina by itself would be the fourth-largest peak in the Cascades, since it is shorter only than two summits on Rainier and the main summit of Shasta. Together, the two summits give Shasta a distinctive saddle-shaped profile (fig. 13.1), making it look very different from the simple conical shape of most Cascade peaks. Most of the surface of Mount Shasta is smooth and unglaciated except for the seven glaciers on the upper slopes of the north and east sides; the largest valley (Avalanche Gulch) no longer has a glacier in it. Its glaciers are larger than those that remain in the Sierras. Shasta has the longest glacier in California (Whitney Glacier) and the most voluminous (Hotlum Glacier). These glaciers are further evidence that there is no tunnel city or hangar door on the mountain, since if the UFO tunnels were truly ancient, the feature would have been destroyed by glaciers. If it were produced after the glaciers retreated, it would be obvious to the many geologists who have hiked up and down these glacial valleys studying them.

The New Age accounts often say that Shasta is five hundred million years old, or even older. In fact, it is a very young volcano and one of the youngest in California. Almost no parts of California existed five hundred million years ago in the Cambrian Period except for a few places in the

Mojave Desert. The oldest dated deposits near the volcano tell us that Shasta began to form only half a million years ago. About three hundred thousand years ago, the entire north side of the mountain collapsed, sending a debris avalanche with about 27 square kilometers (6.5 cubic miles) of material about 45 kilometers (28 miles) down into the Shasta Valley; it has since been deeply eroded by the Shasta River. The oldest dated cone is Sargents Ridge on the south side, which was an andesite-dacite eruption dating to about one hundred thousand years ago. These ancient flows were then glaciated as recently as twenty thousand years ago.

About 20,000 years ago, another eruption occurred, forming Misery Hill, a subsummit peak just south of the main summit. The next eruption was Shastina itself about 9,800 years ago, sending flows that reached as far as Black Butte, 11 kilometers (7 miles away). Since Shastina formed after the last Ice Age, it is unglaciated and has no glacier valleys. Then came the eruption of Hotlum Cone, about 8,000 years ago, producing the 150-meter- (500-foot-) thick Military Pass flow that roared as far as 9 kilometers (5.5 miles) down the northeast face. A dacite dome then grew in Hotlum Cone, which has erupted nine times since the initial eruption 8,000 years ago, producing at least four big hot volcanic mudflows (called lahars) that rumbled 12.1 kilometers (7.5 miles) down from the summit. Its most recent eruption was witnessed by the French explorer La Perouse who was sailing off the coast of California in 1786. Shasta's eruptions are spaced roughly 600 to 800 years apart, and the volcano has been dormant since 1786. While it is apparently not overdue for an event, it is certainly capable of a catastrophic eruption at any time.

There is a huge mythology about Shasta, but none of it withstands the scrutiny of scientific investigation. Instead, we find a complex but well-understood volcano, with no evidence of tunnels, underground cities, or UFO hangar doors. The majority of the UFOs seen over Shasta can be attributed to lenticular clouds. The legends, myths, and bizarre tales that entice so many New Agers and UFO cultists to Shasta may be fascinating (and a boon to their tourist industry), but they are pure fantasy. There is no scientific evidence for any of them.

⑭

The Myth of Atlantis

A Footnote to Plato

Perhaps the most famous mythical place in all of the fine arts, literature, and pop culture is the lost island/continent of Atlantis. It is found in hundreds of literary works, poems, musical compositions, artworks, paranormal and New Age literature, and many science-fiction movies and comic books. One of the hottest films of 2018, *Aquaman*, has scenes set in Atlantis, and it appeared in the 2017 film *Wonder Woman* as well. Atlantis is so familiar as an icon of a lost kingdom or sunken world that just its name evokes different visions of lost peoples of the ancient past or cities of mermaids beneath the waves.

The source of the Atlantis myth is well established; it can be found in two of Plato's famous dialogues, *Timaeus* and *Critias*, written about 360 BCE. These dialogues recount the tale of the ancient lawgiver, Solon of Athens, who heard the story when visiting Egypt around 580–590 BCE. In context, the discussion is about what would make a perfect society (which Plato discussed at greater length in *The Republic*). Socrates asks the people around him (his students and friends, including Plato) if they can think of an example of a perfect society. They agree that their own Athens most closely approached a perfect society, while Critias offers Atlantis as the opposite of Plato's perfect society (he appears as a character in both dialogues, *Timaeus* and *Critias*). As Plato (in the words of the character Critias) described it in the *Timaeus*:

> For it is related in our records how once upon a time your State stayed the course of a mighty host, which, starting from a distant point in the Atlantic ocean, was insolently advancing to attack the whole of Europe, and Asia to boot. For the ocean there was at that time navigable; for in front of the mouth which you Greeks call, as you say, "the pillars of Heracles," there lay an island which was larger than Libya and Asia together; and it was

possible for the travelers of that time to cross from it to the other islands, and from the islands to the whole of the continent over against them which encompasses that veritable ocean. For all that we have here, lying within the mouth of which we speak, is evidently a haven having a narrow entrance; but that yonder is a real ocean, and the land surrounding it may most rightly be called, in the fullest and truest sense, a continent. Now in this island of Atlantis there existed a confederation of kings, of great and marvelous power, which held sway over all the island, and over many other islands also and parts of the continent.[1]

The dialogue *Critias* gives further details about the supposed land of Atlantis. The Greek gods divided the lands of the earth, and Poseidon, the god of the sea, was in charge of Atlantis. It was allegedly larger than Libya and Asia Minor (now Turkey) combined, with mountains on the northern end surrounding a great plain about 555 kilometers (345 miles) across and 370 kilometers (230 miles) long. The dialogue describes elaborates series of moats and tunnels carved around the island by the Atlanteans, along with a canal to the sea that would allow ships to sail to the city in the center of the island. The walls of the city were made of red, white, and black rock quarried from the moats and were covered with brass, tin, and the precious metal orichalcum (a mysterious metal mentioned by the ancients but whose identity is still unknown).

In this account, Atlantis was located beyond the Pillars of Hercules (now the Straits of Gibraltar, and thus somewhere in the Atlantic) at a time when few seafaring cultures of the Mediterranean dared sail out into the Atlantic Ocean. Supposedly, the Atlanteans conquered not only the western Mediterranean but also lands as far east as Egypt. Critias says that nine thousand years before him, or about eleven thousand years ago (and before any human history was recorded), the Atlanteans conquered all of the western Mediterranean and southern Europe and made them all slaves. The Athenians then led a resistance to the Atlanteans and eventually won and liberated the invaded territories. (This alone makes the story mythical, since Athens was not a real city, let alone an empire, earlier than 1400 BCE.)

According to legend, after the Atlanteans were defeated, their island was sunk by an earthquake and became an impassable shoal in the ocean, so no boats could sail near it. In the words of the *Critias*, "But afterwards there occurred violent earthquakes and floods; and in a single day and night of misfortune all your warlike men in a body sank into the earth, and the island of Atlantis in like manner disappeared in the depths of the sea. For which reason the sea in those parts is impassable and impenetrable, because there is a shoal of mud in the way; and this was caused by the subsidence of the island."[2]

Scholars have noted that there was an earlier account of a place named Atlantis by Hellanicus of Lesbos around 450–430 BCE, although only fragments of this story still survive. Plato apparently borrowed a lot of his description of Atlantis from Hellanicus, especially the genealogy of Atlantean kings, which closely follows what we know of Hellanicus's work. The scholar Rodney Castleden says that Plato wrote *Critias* around 359 BCE, after traveling from Sicily back home to Athens.[3] In addition to Castleden, several other scholars note similarities of the ancient description of Syracuse to Plato's Atlantis, so it may have served as an inspiration for the idea.

Since that time, there has been intense scholarly debate as to whether Atlantis was a metaphor or myth used by Plato to illustrate the contrast between the tyrannical empire of Atlantis and the democratic society of Athens or whether the place really existed. Certainly, Plato's accounts are full of stories of Greek gods, impossible dimensions, and odd geography, which makes Atlantis seem all the more mythical. In addition, the supposed time frame of 11000 BCE, before any real civilization existed on earth (the oldest cities on earth, like Jericho, are about this age, but the Greek city-states were no more than 1,500 years old when Plato wrote his dialogues), certainly suggests that it is more legend and myth than a real place. If it was originally based on a real place, Plato's accounts are obviously so embellished and legendary that they don't serve as much of a reliable guide to what the real Atlantis was like—if it existed at all.

From Plato to the Paranormal

In the more than two thousand years since Plato, dozens of known writers and scholars (and many more who were probably lost) have speculated on the legend of Atlantis, where it might be and which parts of Plato's story are legend and which are real. About 450 CE, the medieval Neoplatonist author Proclus wrote a commentary on the *Timaeus*, describing Atlantis:

That an island of such nature and size once existed is evident from what is said by certain authors who investigated the things around the outer sea. For according to them, there were seven islands in that sea in their time, sacred to Persephone, and also three others of enormous size, one of which was sacred to Hades, another to Ammon, and another one between them to Poseidon, the extent of which was a thousand stadia [two hundred kilometers]; and the inhabitants of it—they add—preserved the remembrance from their ancestors of the immeasurably large island of Atlantis which had really existed there and which for many ages had reigned over all islands in the Atlantic sea and which itself had like-wise been sacred to Poseidon.[4]

After the end of the Greek, Hellenistic, and Roman Empires, the legends of Atlantis were picked up by Jewish and Christian scholars all through the Middle Ages, adding many more layers of speculation and myth to what was already a widespread legend. By the 1500s, scholars began to suggest that Atlantis was actually the newly discovered American continents, although clearly these lands had not foundered and sunk into the ocean. Francisco López de Gómara, Francis Bacon, Alexander von Humboldt, and Janus Johannes Bircherod all identified Atlantis with parts of the New World; Bircherod went so far as to say, "*Orbe novo non novo*" ("the New World is not new"). The famous seventeenth-century scholar Athanasius Kircher, a founder of Egyptology and a pioneer in geology, believed Atlantis was real and had been a small continent in the middle of the Atlantic (fig. 14.1).

With the conflation of the Americas with Atlantis, the peoples of the Americas (especially the Mayans, with their impressive ruins that were then poorly understood) were soon taken to be the sources of the myths about the Atlanteans. This even led to early speculation in 1596 by the Flemish geographer Abraham Ortelius that Atlantis sank when continents drifted apart, one of the earliest written suggestions of continental drift ever recorded. In Ortelius's words,

Unless it be a fable, the island of Gadir or Gades [Cadiz] will be the re-maining part of the island of Atlantis or America, which was not sunk (as Plato reports in the Timaeus) so much as torn away from Europe and Africa by earthquakes and flood. . . . The traces of the ruptures are shown by the projections of Europe and Africa and the indentations of America in the parts of the coasts of these three said lands that face each other to anyone who, using a map of the world, carefully considered them. So that

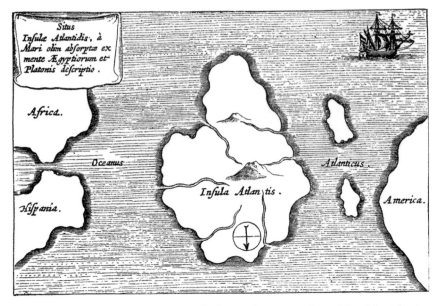

Figure 14.1. Athanasius Kircher's map of Atlantis, placing it in the middle of the Atlantic Ocean, from *Mundus Subterraneus*, published in Amsterdam in 1669. The map is oriented with south at the top. (*Courtesy Wikimedia Commons.*)

anyone may say with Strabo in Book 2, that what Plato says of the island of Atlantis on the authority of Solon is not a figment.[5] Even more prevalent in those centuries were the accounts of pseudohistory that conflated legends with real places, some of which were intended to be taken seriously but others of which were clearly metaphorical works of fiction. For example, Sir Thomas More, after reading about Plato's Atlantis and the writings of early travels in the Americas, was inspired to write *Utopia* (whose name literally means "no place" in Greek). In 1623, Sir Francis Bacon described visions of utopian societies in the Americas, which he called *The New Atlantis*. By the middle of the 1800s, many scholars decided that there was some sort of connection between Atlantis and Mesoamerican cultures like the Mayans. Some of these people, like the pseudoarcheologist Augustus le Plongeon, excavated some of the Mayan ruins and then imagined fantastical connections between the Mayans and Egyptians, which have never been documented, and even more questionable connections between the Greek and Mayan languages.

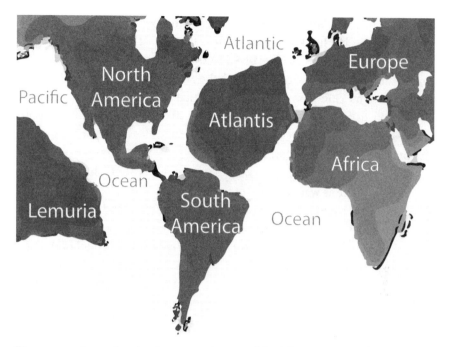

Figure 14.2. A map showing the supposed extent of the Atlantean Empire, from Ignatius L. Donnelly's *Atlantis: the Antediluvian World*, 1882. (*Courtesy Wikimedia Commons.*)

The most influential of these authors was Ignatius Donnelly, whose 1882 book, *Atlantis: The Antediluvian World*, claimed that all known civilizations were descendants of Atlantis (fig. 14.2). He thought that the parallels between the creation myths of the Old World and New World were proof that these myths were real, that the Garden of Eden really existed, and that Atlantis was destroyed by Noah's flood. None of these ideas have borne up under later scholarly scrutiny (the similarities in creation and flood myths are more likely due to common threads in all ancient cultures), but they were hugely influential as the starting point for many different paranormal cults that followed.

The most famous was the cult of theosophy, founded in the 1870s by the Russian mystic Helena Blavatsky and her partner, Henry Steel Olcott. It was the source of many of the modern occult, New Age, and other mystical movements and other weird ideas. Blavatsky began with Donnelly's notions and then dictated her mystical ideas in her book *The Secret Doctrine*, published in 1888. Supposedly translated from documents originally written in Atlantis, her book introduced ideas that the Atlanteans were the first "Root Race," which has since evolved into the fifth and most advanced humans, the "Aryan Race." She claimed the Atlanteans were the good guys, not the villains as in Plato's accounts, and that they had reached their peak civilization between about one million and nine hundred thousand years ago, then were torn apart by the supernatural and psychic powers of the Atlanteans during their internal wars.

Almost anyone hearing *Aryan race* recognizes that phrase, because these occult notions were important to the early mythology of Nazism. They were officially promoted in Germany by Hitler and the Nazi Party as a justification for their "racial superiority" over "inferior races" like Jews, Russians, Poles, Roma, and blacks. Among the major writers of this school of thought was Julius Evola, who in 1934 claimed the Atlanteans were Nordic supermen from "Hyperborea," the old Greek name for Northern Europe, and had originated in the North Pole. Alfred Rosenberg wrote in 1930 of the "Nordic-Atlantean" or "Aryan-Nordic" master race. Ironically, Blavatsky described the Atlanteans as olive-skinned people with Mongoloid traits, supposedly the ancestors of Amerindians, Mongolians, and Malayan peoples.

Finally, the famous psychic Edward Cayce (1877–1945) often did his readings in a trance while speaking about people he was channeling from Atlantis; he told his subjects that they were reincarnations of someone from Atlantis and claimed to have visions of detailed scenes from the lost continent. Among his final predictions were that Atlantis would rise again during the 1960s, leading to a surge in his popularity in the 1960s as occultists and New Age mystics all awaited the great event—which never happened.

As the years have gone by, the idea of Atlantis has faded in popularity among occultists and paranormal believers, although it is more prevalent than ever in pop culture. Modern Plato scholars, with the benefit of a more accurate historical knowledge of the ancient world and his writings, have pointed out that Atlantis is almost certainly a myth or a metaphor. Professor Julia Annas of the University of Arizona wrote,

> The continuing industry of discovering Atlantis illustrates the dangers of reading Plato. For he is clearly using what has become a standard device of fiction—stressing the historicity of an event (and the discovery of hitherto unknown authorities) as an indication that what follows is fiction. The idea

is that we should use the story to examine our ideas of government and power. We have missed the point if instead of thinking about these issues we go off exploring the sea bed. The continuing misunderstanding of Plato as historian here enables us to see why his distrust of imaginative writing is sometimes justified.[6]

Scholar Kenneth Feder reminds us that in the *Timaeus*, Plato provides a major clue. Critias supposedly tells Socrates about his society, "And when you were speaking yesterday about your city and citizens, the tale which I have just been repeating to you came into my mind, and I remarked with astonishment how, by some mysterious coincidence, you agreed in almost every particular with the narrative of Solon."[7] A. E. Taylor wrote about this passage, "We could not be told much more plainly that the whole narrative of Solon's conversation with the priests and his intention of writing the poem about Atlantis are an invention of Plato's fancy."[8]

So Where Was Atlantis?

First, let's get right to the point: does modern science give us any reason to believe that there is a sunken continent beyond the Pillars of Hercules in the Atlantic that was the source of the legend? The simple answer is no. In the 1950s, Marie Tharp and Bruce Heezen used detailed echo-sounding records of dozens of ships to create the first complete map of the Atlantic Ocean floor. There is no sunken island or plateau of any size anywhere in the Atlantic that could fit the bill—just a number of islands and a few small seamounts (nowhere big enough to be Atlantis), and, of course, the giant Mid-Atlantic Ridge down the middle, most of which is more than 3,000 meters (10,000 feet) below sea level. Even more detailed studies of much of the seafloor shows no evidence of sunken civilizations. Ocean sediment cores show that most of the oceanic crust in the Atlantic Ocean is steadily being buried under a gentle rain of mud and the shells of plankton.

Before Tharp and Heezen's mapping and the detailed geology of the Atlantic was known, some pointed to the Canary Islands or the Madeira Islands as a possible source of the Atlantis myth. However, the geology of these islands shows that they have been steadily rising over the past few million years, not sinking into the ocean. The Azores, located in one of the few places where the Mid-Atlantic Ridge stands above the ocean, have also been proposed, but sediment cores around it shows that for millions of years

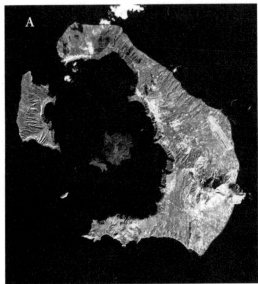

Figure 14.3. The explosion of Santorini, or Thera, caused a catastrophe in the ancient Mediterranean. A. Satellite image of the islands of Santorini. From the Minoan eruption event, this location is one of many sites purported to have been the location of Atlantis. The caldera was originally a ring-shaped feature, but some of the ring was destroyed during its cataclysmic eruption. B. View of the modern Greek towns that hug the steep rim of the caldera. (*Courtesy Wikimedia Commons.*)

it has been an undersea plateau, with no evidence that it once rose above the waves, became civilized, and then sank back into the ocean.

If there is nothing in the Atlantic that could have led to the legend, many scholars have looked at other areas around the Mediterranean and Europe that might have been the seed for the idea, even if not in the right place. One of the best candidates is the eruption of the island of Thera (also called Santorini) in the Aegean Sea (figs. 14.3 and 14.4). Sometime between 1600 and 1627 BCE, Thera exploded in a massive volcanic eruption that blew the island apart and ejected 60 cubic kilometers (14 cubic miles) of rock and ash

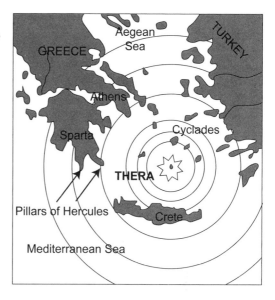

Figure 14.4. Map showing the Aegean region, with Thera, Crete, the Gulf of Laconia, and the Pillars of Hercules. (*Redrawn out of copyright.*)

into the sky. One of the largest volcanic events in history, with a volcanic explosivity index of 7, it erupted four times as much material as the famous 1883 eruption of Krakatau in Indonesia. The ash plume rose into the stratosphere, which spread ash around the world and darkened the skies to cause volcanic winter in places as far off as China, California, and Sweden. Pumice deposits were found as far away as Egypt. Apocalyptic rainstorms pounded Egypt with devastation, as described on the Tempest Stele of Ahmose I.

The biggest effect, however, was giant seismic sea waves or tsunamis (*not* tidal waves, since they have nothing to do with tides) that rose up to 150 meters (492 feet) high as they swept across most of the Greek islands in the Aegean. The biggest waves virtually wiped out the ancient Minoan civilization on the island of Crete, only 110 kilometers (68 miles) away (fig. 14.4). The Minoan civilization was one of the most advanced in the world by that time, rivaled only by the archaic Greek Mycenean culture of the mainland. Minoan culture was famous in the Greek legends of King Minos, who destroyed people by letting them wander in the labyrinth until they were devoured by the Minotaur, a half-man, half-bull beast. According to the myths, Theseus went into the labyrinth and slew the Minotaur and then used a thread from the king's daughter Princess Arachne to find his way out again and claim his victory.

These are just myths, of course, but paintings on the walls of the Palace of Knossos show how important the bull was to Cretan mythology. They held contests where young men would try to vault over the horns of the bull without being killed. There is still a lot of debate among scholars as to whether Thera was the main factor in the fall of the Minoan civilization, although as a seafaring culture, they would have been devastated if all their cities were flooded and destroyed by the tsunamis, all their boats had been washed away, and most of their people killed. Most recent evidence suggests that the eruption weakened Late Minoan I culture, even though it survived the Thera crisis. Then the Minoans were conquered by the Mycenaeans at about the end of the Late Minoan II culture and vanished forever.

Today, the ruins of Knossos are excavated and available for visitors to see, and scientists have documented the fingerprints of the Thera eruption all around the Mediterranean as well as in Greenland ice cores and even tree rings of bristlecone pines in California. Was the ancient Minoan civilization the basis for the legend of Atlantis? After all, it happened over 1,200 years before Plato's account, so it was lost in the mists of time by then. Even today archeologists can't read most of the Minoan texts, since their writing and alphabet have never been fully deciphered. Another consideration was that the mountains on each side of the Gulf of Laconia (the large south-facing gulf on the Peloponnese in southern Greece, which had Sparta at its northern end) had rocks at the mouth of the gulf that were also called the Pillars of Hercules (fig. 14.4) before the sixth century BCE (the Straits of Gibraltar got that name much later). The mouth of the gulf faces toward Crete, so this may be what Plato was thinking of when he situated Atlantis "beyond the Pillars of Hercules." If so, Crete and the Thera eruption make the best candidates for Atlantis, assuming it was real at all.

Another popular candidate is the city of Helike, on the Gulf of Corinth and the north shore of the Peloponnesian Peninsula, first suggested in 1985 by Swiss scholar Adalberto Giovannini.[9] Helike was the capital of the region of Achaea and a rich and powerful city since the days of Homer. It supposedly provided soldiers for Agamemnon's army that conquered Troy. In the winter of 373 BCE, it was utterly destroyed by a huge earthquake and tsunami that drowned the city and sunk it below the waves as the coastal plain on which it had been located sank down into the sea along a fault. This was within the lifetime of people that Plato knew, so the event was relatively recent but may have inspired Plato to use it for the model of his mythical ancient Atlantis from thousands of years in the past. The remains of the city were a popular tourist destination in ancient times, since people could float over the sunken ruins in boats and even see the statue of Poseidon and his seahorses deep beneath the surface. Many ancient scholars recorded their visit to the sunken city, including the Greek geographers Strabo, Pausanias and Diodoros of Sicily, and Romans like Ovid and Aelian. Through the Roman period, visitors would sail over the city to see the ruins, although reports suggest that it was becoming more and more difficult to see. Eventually, it was silted over and the location of the city was lost to memory, only to be rediscovered in 2001 sunken into an ancient lagoon near the modern town of Rizomylos. In recent times, the history of Helike has become much better known as archeologists have excavated much of the city and studied it in detail.[10]

Other possibilities have been proposed as well. When sea level was at its lowest during the last peak glacial around twenty thousand to eighteen thousand years ago, much of the floor of the North Sea was exposed as a broad floodplain called Doggerland, which connected Britain to Northern Europe. When sea levels rose around ten thousand years ago during the current interglacial, most of that region was flooded. Archeologists have dredged numerous artifacts from the floor of the North Sea to show that there were once a lot of people and cities on Doggerland, who were then pushed away when the rise of sea level drowned the region.

Yet another idea is that there are flooded marshlands and drowned shelf areas in the Andalusia province of Spain, between Cadiz and Seville provinces, and just west of Gibraltar, that might have been the source of Atlantis. This idea is very controversial. Just like Lemuria, people have placed Atlantis all over the world, including many places far from Greece or even the Atlantic Ocean. These places include Antarctica, Cuba, the Bahamas, India, Indonesia, and even the Bermuda Triangle. The connections to the occult, paranormal, and pseudohistory are never far away when people start talking about Atlantis!

(15)

The Mysterious Ley Lines

Lines on a Map

One of the more obscure but nevertheless bizarre ideas about the earth is that there is a network of lines called *ley lines* that connect up points on a map. The lines allegedly have some sort of meaning, usually paranormal or supernatural. They have entered the vocabulary of the occult and paranormal and New Age believers and are appearing in all sorts of places, including the 2016 remake of the movie *Ghostbusters* (with the all-female ghostbusting team). It is a sure sign that a fringe idea has reached the mainstream when it completely loses its original meaning, becomes part of the pop culture, and is associated with other popular ideas about the supernatural. The dialogue in the movie has the female Ghostbusters looking at a map and discovering ley lines connecting places all over Manhattan. In their words, they are a "hidden network of energy lines that runs across the earth" and "a current of supernatural energy." Where the ley lines intersect "it's an unusually powerful spot."

In many cultures, straight lines were often given great significance, because nature doesn't often arrange features in straight lines. In Europe, straight-line paths were associated with spirits or fairies, so there were all sorts of legends associated with fairy roads. (Circles of sprouting mushrooms were called fairy rings, but they are simply due to the fact that the fungus that makes the mushrooms grow equidistant from a central point.) The street where bodies were carried from the town to the graveyard was called the corpse road (also known as the bier road, coffin line, coffin road, funeral road, lych or lyke way, or procession road), and people avoided it.[1]

In other parts of the world, long straight ceremonial paths are often visible from the ground and even from the air. For that reason, UFO believers make a big fuss about anything running straight along the ground, even when it has a simple natural explanation. The famous 1968 book *Chariots of the Gods?: Unsolved Mysteries of the Past* by Erich von Däniken claimed that long straight lines visible from the air were landing strips for UFOs or signals to aliens. These ranged from the Nazca lines in Peru to the cart ruts in Malta, and similar linear features in Australia, Saudi Arabia, and the Aral Sea. French UFO author Aimé Michel published *Flying Saucers and the Straight-Line Mystery* in 1958, where he claimed that UFO sightings were found along lines that stretched a long distance across the grid.[2] He called these features "orthotetic lines," and they are a big part of the UFO belief system now.

Naturally, Atlantis was tied into these ideas, as it features in nearly every occult and paranormal story. In 1969, British author John Michell published a book, *The View over Atlantis*, which resurrected the old notion of ley lines, and they soon became part of the New Age beliefs of the counterculture. Michell was famous for originating many of the ideas of the New Age and earth mysteries movements. He was obsessed with crop circles as proof of alien visits, and he even compiled a book of Adolf Hitler quotations. He thought that ancient human traditions were all connected to the divine but that this connection had been lost due to modernity. He believed the old traditions would come back and humans would enter a golden age, centered in Britain.

The idea of linear connections and patterns is widespread in human culture. The now-faddish Chinese art of feng shui claims that the alignment of objects (like furniture) in a room creates a mystic "balance" or "harmony." Linear features were often claimed to be the "arteries" of the earth or associated with "veins" of energy flow, similar to the Chinese idea of chi, or qi, pathways in the body supposedly tapped during acupuncture. From this, it was only natural that the rings of standing stones (like Stonehenge) would be compared to the acupuncture needles.

As geologist and skeptic Sharon Hill pointed out,

> It was a small jump to connect genuine scientific concepts of magnetic anomalies and telluric current to sciencey-sounding "earth energy lines." ... Some suggested that the power in ley lines drew from the telluric current and that such power could be manipulated or used for societal advantages. Watkins did not think ley lines could be found with dowsing rods, but this creative subversion of his ideas occurred rather smoothly. Author Guy Underwood pushed the dowsing craze associated with ley lines in the 1970s in the midst of the New Age wave of magical earth ideas. ...

So, in the 1970s, ley lines were seen as channels of mystical "energy." We can measure energy, yet no one was successful in measuring this particular energy scientifically, though. Ley proponents said the energy was too subtle or that research on the lines should be funded because of the potential to humankind. Locations associated with ley lines became more magical as New Age popularity increased—places like Glastonbury . . . and Stonehenge. Giants were said to be buried in the barrows as they were associated with incredible ley line power. This power was said to even cause the stones to come alive and move. (Another testable claim where evidence was never provided.)

Crossed lines have always been associated with magic. Crossroads are particularly notable as supernatural areas. Where ley lines supposedly crossed, a strong "power spot" was created. These nodes could produce an energy "vortex." Vortexes (vortices) are handy devices to explain an area as particularly prone to strange phenomena. No evidence for such energy vortexes exists, either.

Ley lines as plot devices or explanations in fantasy fiction media began in the 1960s with authors like Thomas Pynchon and continued with concepts of conspiracy ideas of secret knowledge about sacred geometry of the earth. And, of course, we have ley lines and the vortex as key plot device in the new Ghostbusters, which brings us to the paranormal community connection—the intersection where supernatural ideas and vague scientific concepts crash and merge.[3]

In more recent years, paranormal research has gone into high gear, with television shows and paranormal websites and books connecting features on the map. According to Sharon Hill,

Ley lines as sources of some undefined kind of energy were picked up as potential explanations for the manifestation of psychic energy that ghost researchers assumed was real. Genuine magnetic anomalies were turned into magical areas and connected to paranormal reports. Often, the researchers never actually checked for historic ley lines or geological anomalies, they just assumed they were there because it was a very sciencey conclusion to make that sounded plausible (and the public accepted it). Use of dowsing rods to find this psychic energy is also common for paranormalists.

192

Ley nodes were commonly attributed to areas of "high strangeness"—a higher than normal occurrence of weird things like anomalous lights, poltergeist activity, bizarre creature encounters, and UFO sightings.[4]

Hill cites this passage from *Supernatural* magazine as an example of how far ley lines have drifted into the realm of the spooky and bizarre and supernatural:

> Well most of the Earths leys are positive but when two of these leys cross or intersect a vortex of negative energy is then created. It is like a powerful magnet attracting all kinds of lower vibrational spirit, energy or entity and even sometimes people. These entities can then draw off the energy, feed on it and use it to manifest. Bodmin Jail (Cornwall) is a place where two such energy lines cross and therefore they form lower energy vortexes and this, in turn, will also affect the way people behave in such places. They will be prone or influenced to lower vibrational thoughts, paranoia, anger, ego and fear etc. It can be a source of food to an entity to recharge their essence.[5]

Thus, the concepts of ley lines and linear features in general have veered far off into the realm of the occult and paranormal. But what was the original source of these ideas?

Ley of the Land

Where did the idea of ley lines come from? Surprisingly, it originated from the legitimate research of a self-taught antiquarian and amateur archeologist named Alfred Watkins, who was curious about how many ancient sacred sites in Britain seemed to form lines on a map. According to his own account,[6] on June 30, 1921, Watkins was visiting Blackwardine in Herefordshire, driving along a road near the now-vanished village. He was on his way to see the archeological investigations of a Roman camp nearby and stopped by the road to consult with his worn map, which was already marked with many important ancient sites: standing stones, hill forts, wayside crosses, causeways, and ancient churches on mounds. He suddenly noticed that many of the features formed a straight line on his map "like a chain of fairy lights," in the words of his son.[7] Later, standing on one of the ancient mounds, he noticed that

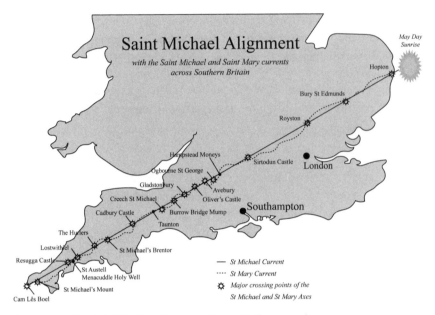

Figure 15.1. The St. Michael's Alignment. (*From Watkins, 1922.*)

many of the footpaths seemed to connect one mound to another in a straight line (fig. 15.1). In 1922, he published a paper on the idea titled "Earth British Trackways" and then expanded on the notion in a 1925 book, *The Old Straight Track*, and a 1927 book, *The Ley Hunter's Manual*.[8]

Watkins coined the term *ley line* because he noticed that the syllable *ley* frequently occurred in the place-names along the lines. In his opinion, *ley* was an ancient word for "trackway" that was incorporated into the name. Supposedly these straight-line paths were surveyed by prehistoric dodmen who tried to find the straightest paths through the once densely forested landscape of Britain. These paths often began and ended at high spots on the landscape that served as navigation points, especially if they were important castles or churches. Later archeologists, aided by almost a century more of detailed research not available in 1921, have disputed Watkins's interpretations of the archeology of Britain. In addition, philologists have dismissed the appearance of *ley* in place-names as a pure coincidence. Nevertheless, for a while in the 1920s and 1930s, Watkins's ideas were very popular.

Watkins was not the first to notice the linearity of features in Britain. In 1882, G. H. Piper presented a paper at the Woolhope Club, where he

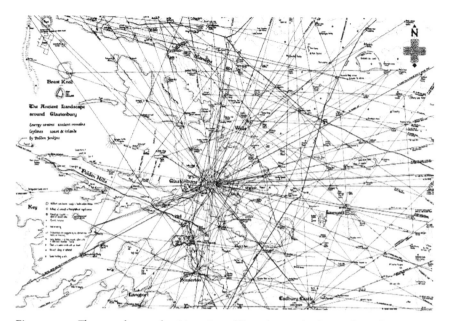

Figure 15.2. The complex spaghetti junction of lines that can be drawn through Glastonbury. (*From Michell, 1969.*)

noted that "a line drawn from the Skirrid-fawr mountain northwards to Arthur's Stone would pass over the camp and southernmost point of Hatterall Hill, Oldcastle, Longtown Castle, and Urishay and Snodhill castles."[9] Watkins may have also recalled reading a presentation given by William Henry Black to the British Archeological Association in Hereford in September 1870, titled "Boundaries and Landmarks." In that talk, Black suggested that "monuments exist marking grand geometrical lines which cover the whole of Western Europe."[10]

Through the rest of his life, Watkins kept finding more and more linear connections through ancient British sites. For example, he claimed that St. Ann's Well in Worcestershire is the beginning of a ley line that runs through St. Pewtress Well, Walms Well, and Holy Well along a ridge in the Malvern Hills. Paul Devereux, part of Watkins's archeological community, drew a Malvern Ley, a ten-mile line passing through Pauntley, Redmarley D'Abitot, Whiteleaved Oak, Midsummer Hill, Shire Ditch, Wyche Cutting, and St. Ann's Well.[11] However, by the 1930s, the study of ley lines had fallen into obscurity.

They were revived in 1969 by John Michell, who claimed there was a "Circle of Perpetual Choirs" with the central point of Whiteleaved Oak, located about the same distance from many historic sites, including Llantwit Major, Goring-on-Thames, Stonehenge, and Glastonbury (fig. 15.2).[12] Located in central Somerset, Glastonbury is often associated with sightings of UFOs and with many other myths and legends, including Joseph of Arimathea, the Holy Grail, and King Arthur. In medieval times, there was a legend that Joseph of Arimathea, who provided the tomb in which Jesus was buried, allegedly took the Holy Grail to Britain. Many of the quests for the Holy Grail focused on the site of Glastonbury, where he supposedly hid it.

In addition, about 1190, Gerald of Wales wrote that Arthur's legendary place known as Avalon was the region around Glastonbury. The monks of Glastonbury Abbey claimed to have found the bodies of King Arthur and Queen Guinevere. According to his account,

> What is now known as Glastonbury was, in ancient times, called the Isle of Avalon. It is virtually an island, for it is completely surrounded by marshlands. In Welsh it is called Ynys Afallach, which means the Island of Apples and this fruit once grew in great abundance. After the Battle of Camlann, a noblewoman called Morgan, later the ruler and patroness of these parts as well as being a close blood-relation of King Arthur, carried him off to the island, now known as Glastonbury, so that his wounds could be cared for. Years ago the district had also been called Ynys Gutrin in Welsh, that is the Island of Glass, and from these words the invading Saxons later coined the place-name "Glastingebury."[13]

For generations, Britons took these stories seriously, and in 1278 King Edward I (known as "Longshanks," conqueror of Wales and Scotland, and villain of Mel Gibson's William Wallace movie *Braveheart*) had the skeletons reburied at Glastonbury Abbey with high ceremony.

Although this version of the Arthurian legend was convincing in the Middle Ages, today it is regarded as pseudoarcheology. Most modern historians regard this as a publicity stunt by the monks of Glastonbury Abbey to raise funds to repair the abbey. Nevertheless, the legend motivated generations of pilgrims to visit Glastonbury Abbey until the Reformation, when Henry VIII broke away from the Catholic Church. Many later writers connected the story of Arthur in Glastonbury with the legendary visit of Joseph of Arimathea to bring the Holy Grail to England. Allegedly, Joseph

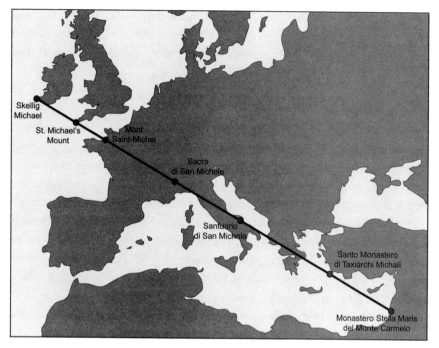

Figure 15.3. St. Michael's Line also crosses from Ireland to the Holy Land. (*Redrawn from several sources.*)

stuck his staff in the ground, where it bloomed into the legendary Thorn of Glastonbury (a tree that survived until 1991). Others connected Avalon and Glastonbury with earth mysteries and even the myth of Atlantis. Today, there are many mythological tales and even modern romances about Avalon, such as *The Mists of Avalon, A Glastonbury Romance,* and *The Bones of Avalon.* Glastonbury is the location of many New Age festivals every summer, where believers in the mysterious and occult congregate and celebrate earth myths.

Since Watkins's original work, many others have drawn ley lines all over Europe. For example, St. Michael's Line connects a series of sites that are named after St. Michael from the Holy Land to Ireland (fig. 15.3). There is almost no place on earth that has not seen ley lines drawn on it, except possibly Antarctica.

Certainly, there are linear features that ancient Britons used and considered sacred. For example, Stonehenge and many other standing stones are often arranged so that they line up with the rising or setting sun during winter solstice and sometimes summer solstice. The ancient pathway that led straight to Stonehenge is also aligned with astronomical landmarks. But are all the other ley lines real?

So Why the Linear Features?

All of these ideas reflect a common tendency for humans to see patterns where there are none, a phenomenon known as pareidolia, or what Michael Shermer called "patternicity."[14] We look at clouds and see things, even when we know that they are just random patterns in water vapor trapped in the upper atmosphere. The famous Rorschach tests are just random blots of ink on pieces of cardboard, but psychologists still use them today to understand a patient's state of mind. Our brains are so heavily programmed this way that people will see the Virgin Mary or Jesus in a grilled cheese sandwich, or the face of Jesus in a whole range of objects, and other such ideas. Especially in Catholic parts of the world, the brain is so heavily bombarded by images of Jesus and Mary that almost any random pattern of light and dark can be shoehorned into recognition of these familiar images.

In the case of ley lines, archeologists were already skeptical of Watkins's work as soon as it was published. Some of them considered his ideas so fantastic and far-fetched that they rejected the entire notion out of hand. Archeologist O. G. S. Crawford even went so far as to refuse to allow ads for Watkins's book *The Old Straight Track* in the leading journal *Antiquity*. The main problem with the ley line hypothesis is that there are so many archeological sites in Britain, an area with so much history and such a dense population, that almost any line would connect quite a few just due to coincidence.

Ever since the 1980s, a number of rigorous mathematical studies have shown that this alignment could be purely due to chance or coincidence. One study concluded that "the density of archaeological sites in the British landscape is so great that a line drawn through virtually anywhere will 'clip' a number of sites."[15] A 1989 study by David G. Kendall, using the mathematical technique of shape analysis, found that the occurrence of straight lines on the map of standing stones in Britain could be due to random chance.[16] Archeologist Richard Atkinson took a map of telephone boxes in London and generated a series of "telephone box leys," once again pointing out that any dense set of points can be connected with lines, whether or not those lines have any meaning.[17] Another study showed that you can take a map of

Figure 15.4. The plot of pizza
restaurants in London is so dense
with points that almost any
line could be drawn to fit them.
(*Redrawn out of copyright.*)

pizza restaurants in London and find any number of lines that connect eight
or more of them (fig. 15.4).[18]

A 2010 article by mathematician Matt Parker from the *Times Online*
(responding to a study by Tom Brooks that claimed geometric patterns in
prehistoric sites in England) deftly satires the entire idea:

> The landscape of England is scarred by history, with landmarks that stretch
> back thousands of years. These prehistoric monuments provide some of
> our only links to previous civilisations, such as the Uffington White Horse,
> the infamous stones of Stonehenge and of course, the ancient Woolworths
> stores. It was these Woolworths sites from a bygone age of cheap kitchen
> accessories and discount CDs that I analysed to try and learn more about
> our ancestral hunter-pick'n'mixers. . . .
>
> Using the locations of the 800 ancient Woolworths stores as my data,
> I found that they also followed precise geometrical patterns with the same
> level of accuracy.
>
> For example, three Woolworths sites around Birmingham form an
> exact equilateral triangle (Wolverhampton, Lichfield and Birmingham
> stores) and if the base of the triangle is extended, it forms a 173.8 mile
> line linking the Conway and Luton stores. Despite the 173.8 mile distance
> involved, the Conway Woolworths store is only 40 feet off the exact line

and the Luton site is within 30 feet. All four stores align with an accuracy of 0.05 per cent.

One possible conclusion from this pinpoint accuracy is that the Woolworths tribal duty managers positioned the stores as a form of "landmark satnav." This allowed travellers to find their nearest outlet for sweets that could be acquired in any combination they desired. This could offer us a fascinating insight into what life was like in 2008 England, and we can't rule out that alien help was required to position stores this precisely and to offer the Ladybird clothing range at such low prices.

Or it could be that I just skipped over the vast majority of the Woolworths locations and only chose the few that happened to line-up. From 800 stores, there are over 85 million possible triangles; the 1500 prehistoric sites that Brooks used give over 561 million triangles. From these millions of options it is easy to pick out the few that seem to be impossibly precise.

It is mathematically known that if you have a sufficiently large set of random data, you can find any pattern that you want with any given level of accuracy. What Brooks and I have discovered says less about any meaning to the patterns and more about how the locations follow a truly random distribution.[19]

Thus, ley lines are no more than products of the human imagination, and the tendency to see patterns in anything, even an assemblage of random dots on a map.

16

Crystal Con Artists

The Beauty of Crystals

Crystals and gems are among the most beautiful objects in nature, and people are often fascinated with them and treasure them (fig. 16.1). They are seductive in their brilliance, their luster, their shine, and often their color and transparency. Many are worth a lot of money to collectors, especially if they are among the groups of crystals and minerals that can be cut and polished to make gems. Some gems, like rare diamonds, are worth enormous amounts of money. Diamonds are among the most valuable of minerals and often are used by people as a hedge against the devaluation of other currencies (although their largest commercial market is for lower-quality industrial diamonds, used for drill bits digging for oil). Diamonds typically go for about $15,000 per carat, and more if the individual diamond has a rare color or quality or is unusually large. On April 4, 2017, the Pink Star diamond, a 59.6-carat pink gem (originally cut down from a 132.5 rough diamond), sold for a record-breaking $71.2 million, making it the most valuable gem ever sold to that point. That's about $120,000 per carat!

Collecting and owning gems and crystals is a popular activity, and the market for gems in particular is an ancient and very lucrative business. Many geologists first discovered their love for the earth when they followed a passion for crystals and minerals by taking a mineralogy class and learned all about the thousands of different minerals that make up the earth. But there is another side to collecting crystals and gems—the pseudoscience of "crystal healing" and the market that sells fake "cures" through "magically" touching minerals and crystals. This is one place where an interest in crystals and minerals goes from harmless fun to harmful in that it often costs enormous amounts for worthless cures. These not only rob people of their hard-earned money for no reason but often prevent them from seeking actual medical care that might have helped. This is definitely one place where weird ideas about earth materials can be dangerous or harmful.

Figure 16.1. Photograph of giant gypsum crystals in the famous Cave of the Crystals, in the Naica Mountains, Chihuahua, Mexico. Under the right conditions of slow, continuous growth, crystals can grow very large. (*Courtesy Wikimedia Commons.*)

What Is a Mineral?

The word *mineral* has all sorts of casual and inconsistent meanings in popular culture, but to geologists and chemists, it has a very strict and clear definition. A mineral:

- Is naturally occurring
- Is inorganic
- Has a definite crystalline structure
- Has a definite chemical composition
- Demonstrates characteristic physical properties

Let's discuss each of these components.

Naturally occurring: There are lots of complex chemical compounds in the world, but if they are not produced naturally, they are not minerals. Thus, a synthetic diamond produced in a lab has all the properties of a diamond mined from the earth, but it's technically not a mineral. Ice formed as

Figure 16.2. Some examples of minerals with a cubic lattice and cubic cleavage. A. A large cubic crystal of salt (the mineral halite) in front of a ball-and-stick model of the cubic lattice of its atoms. (*Photo by the author.*) B. Galena, or lead sulfide, has the same cubic lattice as halite and breaks into cubic crystals with 90° cleavage angles. (*Courtesy Wikimedia Commons.*)

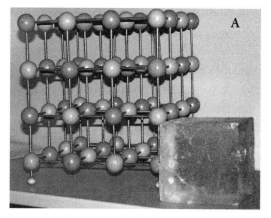

ice crystals or snowflakes are minerals, but not ice in your ice cubes. Most of the stuff sold in a health food store that is called mineral was produced synthetically and, therefore, is not a mineral as a scientist uses the word.

Inorganic: Organic chemicals are built of the element carbon, so most minerals are not made of carbon. However, as there are a handful of important minerals with carbon (such as the mineral calcite, or $CaCO_3$), we'll use *organic* in this context to mean compounds with carbon-hydrogen bonds. Thus, sugar forms beautiful crystals, but it is organic and therefore not mineral. Many of the minerals sold in a health food store are organic and thus not minerals.

Definite crystalline structure: Like the word *mineral*, the word *crystal* has a different meaning to a scientist than it does in popular culture. Typically, people use the word to describe anything that sparkles. In a scientific definition, a crystal must have a *regular three-dimensional arrangement of atoms* in its internal structure, which repeats over and over again. This three-dimensional array is called a lattice. It is analogous to the regular repeated pattern in wallpaper. For example, the atoms of the salt (sodium chloride, or NaCl) crystal (fig. 16.2A) are arranged in a cubic pattern, with each atom of sodium or chlorine forming 90° angles with the others. The same lattice is found in the mineral galena (fig. 16.2B), which is made of lead and sulfur in equal amounts (lead sulfide, or PbS). All minerals have a regular three-dimensional lattice of some kind, often very complex and with many other angles between atoms besides the 90° seen in the simple cubic lattice.

Some things in nature may have a three-dimensional arrangement of atoms, but they are not regular and repeating. Take, for example, volcanic glass, or obsidian (fig. 16.3). At the molecular level, their atoms are not in any kind of repeated pattern; they are in a random tangle of long chains, like a bowl of spaghetti. Thus, a glass is by definition not a crystal. A popular item at many gift shops is "cut glass crystal" drinking goblets and chandeliers, but this is not the definition of crystal that scientists use.

Definite chemical composition: Most minerals have a simple chemical formula, like most other compounds. However, a substitution is allowed if you replace one ion with another of a similar charge. For example, in the mineral calcite (calcium carbonate, or $CaCO_3$), a certain percentage of magnesium can replace the calcium sites in its lattice and still be calcite. However, if it gets to be a fifty-fifty ratio of calcium to magnesium, then it's no longer calcite but a different mineral, dolomite.

Characteristic physical properties: Most of the features of the minerals we have discussed occur at the atomic level. To identify the mineral, you need to know what physical properties are typical of a hand sample of the mineral. These include what color it is, how hard it is (from soft minerals like talc and gypsum to the hardest mineral, diamond), and whether it fractures with an irregular surface or has a cleavage and breaks into many parallel planes, as well as less commonly used properties like density (lead sulfide or galena, for example, is unusually dense because it contains lead), reaction to acid (the mineral calcite fizzes in dilute hydrochloric acid), and magnetism (the mineral magnetite is naturally magnetic).

Many of these mineral properties can be understood by knowing the crystal lattice. For example, the cubic lattice of minerals like salt or halite (NaCl) or galena (PbS) is demonstrated at the hand sample level, since any

Figure 16.3. Obsidian is a form of volcanic glass where the magma has cooled so quickly that no minerals had time to crystallize. (*Courtesy Wikimedia Commons.*)

time you hit and break a piece of these minerals, they will naturally cleave to form cubic faces with 90° angles (fig. 16.2).

Atomic-level properties and crystal lattice can make a big difference in the behavior of a mineral at the macroscopic level. Let's take as an example the two common minerals formed of pure carbon: diamond and graphite. One is the hardest substance in nature, and the other is one of the softest, yet they are chemically identical. Why are they so different? Diamond has a crystal lattice (fig. 16.4) with all the atoms of carbon tightly bonded together and very short, strong chemical bonds. This structure will survive huge amounts of pressure, and an expert diamond cutter has to know exactly how to cleave a large stone into several smaller ones. If the cutter misses, the diamond is ruined. In graphite (the "lead" in a pencil), on the other hand, its carbon atoms are arranged in sheets, with very long, weak molecular

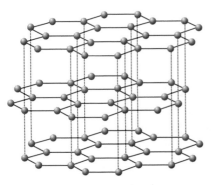 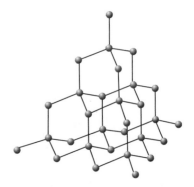

Graphite (solid lines are strong covalent bonds,   Diamond (all bonds are strong covalent bonds)
dotted lines are weaker inter-layer bonds)

Figure 16.4. Comparison of the lattice of diamond and graphite, two different minerals made of the same chemistry, pure carbon. The bonds in diamond are very short and strong, resulting it its extraordinary hardness. Graphite is arranged in sheets of carbon atoms with long, weak bonds between the sheets, so it cleaves easily with pressure. (*Courtesy Wikimedia Commons.*)

bonds between the sheets. Pushing the graphite tip of a pencil across paper is enough to break those weak bonds, leaving tiny flakes of graphite behind on the paper as pencil markings.

Another example is calcium carbonate, or $CaCO_3$, in two different mineral lattices: calcite (the common mineral in limestones and marbles) and aragonite (also known as mother of pearl, or nacre). They are the same chemistry, but their lattices are very different and lead to very different properties. The most obvious of these is that aragonite is much more soluble than calcite, so in weakly acidic conditions, aragonite will dissolve, whereas calcite won't. This is why if you own pearls, it is important to wash your acidic sweat off them after wearing them. For some minerals, knowing the chemical composition is not enough; the crystal lattice makes a huge difference in the properties of the mineral.

Thus, we will stick to the strict scientific meanings of the words *mineral* and *crystal* rather than the common use of the word. To a geologist or chemist, a crystal ball is normally made of glass and thus not crystalline. On rare occasions, it is carved out of a quartz crystal, so in that case it would really be a crystal ball. The shiny glass in a chandelier, or in your crystal glassware, is glass, not a true crystal. Most of the crystals sold in rock shops and New

Age stores are typically single crystals of common minerals like quartz, so they are true crystals. And most of the stuff sold in health food stores are not in fact minerals, even if they claim to be on the label.

Crystal Healing

Many prescientific cultures had legends and beliefs about the magical powers of crystals. This is especially true of many Native American cultures, but they are also found in Asian cultures tied in with the concept of life energy or chi (now spelled *qi*); the Indian practice of reiki; and the Hindu and Buddhist idea of chakras, vortices of life energy thought to connect the physical and supernatural elements in your body. In recent years, the ancient folkloric treatment of gems and crystals as having magical powers has spread to the people engaged in New Age thinking. It is a big part of what is often called "complementary and alternative medicine" (in other words, medicine that has not been subjected to scientific testing to prove that it actually works).

Lots of people think that if a crystal is beautiful and has an almost supernatural look and feel to it, it must have special powers in some way. But a century of careful scientific analysis shows that crystals are nothing more than atoms aligned in a very specific lattice. These crystal healers make claims that crystals do everything from healing to protection, focusing energy, directing energy, and aligning energy, in addition to even wilder assertions. A typical statement from one of the many sites promoting crystal healing reads as follows:

> Research into the structure of the atom over the last few hundred years has revealed that everything in our entire universe is made up of energy. Even solid objects, like a piece of furniture or the hair on your head, are really just vibrations of energy at the most fundamental levels. It may not look like it to your eye, but healing crystals and the cells in your body are made up of the same kind of energy.
>
> Scientists have already figured out how to use the energy inherent in crystals for all kinds of things like keeping time using small quartz crystals in your watch or creating the electronic components to your computer and smartphone. Whether you realize it or not, the energetic properties of healing crystals and stones are widely used in our modern technology.
>
> We even use crystals in our medications. Many pharmaceuticals are made by grinding up minerals that form inside of healing crystals. Even

though our culture has several uses for the energetic properties of crystals, we have neglected to standardize their use in energetic healing.

Just like magnets use energy to attract or repel, healing stones crystals use energy in the same way. When you place certain crystals over certain parts of your body, your energy transforms, vibrates, pulses, moves and shifts in accordance with the properties and energetic signature of the crystal.[1]

Let's unpack some of the myths and misconceptions in this amazing statement. First is a misuse of physics to say that "the entire universe is made of energy." This is, of course, restating Einstein's famous formula, $E = mc^2$, where matter (m) and energy (E) are related. However, this kind of relationship is only true of processes operating at the molecular and subatomic level or of processes happening on the scale of nuclear reactions, stars, and galaxies. It is not operative on the scale of human bodies, so no crystal has the ability to project "energy fields" that can affect human tissue.

Second is the common misconception that the piezoelectric effect of a quartz crystal in a watch is somehow producing energy. No, the energy of your quartz watch comes from an ordinary watch battery, which sends a charge through a tiny quartz crystal under mechanical stress to make it vibrate at a high frequency and thus keep the electronics of your watch moving. That piezoelectric quartz crystal is not emitting any mystical energy, nor is it capable of sending energy large distances. And the electronic components in your technology are not the big quartz crystals that the healers sell but tiny wafers of silicon produced commercially, often not even fully crystalline. They contain information because they have been etched by a laser and have had conducting metals bonded to them in tiny silicon wafers called chips, not because the quartz crystal lattice itself has any information beyond how to grow a larger quartz crystal.

Sure, we use ground-up crystals in our medications, just like we use tiny crystals of sand in a sandbox (crystals of quartz sand less than 2 mm in diameter) or tiny crystals in our potting soil (composed partially of tiny crystals of quartz sand plus the even tinier layered crystals of clay minerals, commonly known as mud). In the case of pharmaceuticals, crystals of inert compounds are often ground up to make the pill, although many of the active organic ingredients may also have a crystalline form. For that matter, putting sugar in your tea or coffee is adding crystals, but it doesn't suddenly make the drink magical, just sweeter. Just because mineral grains and crystals are found all sorts of places doesn't mean they have magical powers.

Finally, the comparison to magnets is completely false. Magnetic minerals are among the few substances that actually have an electromagnetic field around them, but most magnetite crystals have a tiny field that cannot be experienced more than a centimeter away. No other magnetic minerals, like magnetite and hematite, are shown to have *any* kind of energy field that humans could even feel, let alone that affect humans or their mood or tissues.

Another example of the mumbo jumbo found on the crystal healing sites says the following:

> Since healing crystals absorb, attract and repel certain types of energy, it's important to keep your crystals clean. If you use crystals to absorb negative energy, you'll want to get rid of that energy before using them again. Think of this like using a sponge to soak up dirty water. If you want to keep using the sponge, you'll need to squeeze out the dirty water and clean it up so the next plate you wash doesn't also get dirty.
>
> When you purchase healing crystals or stones in a store or online, they have been absorbing and repelling the energy of everyone who has touched them. Before you use them on yourself, you'll need to cleanse their energy and align it with yours.
>
> Doing this is simple. You can soak your crystals (don't soak selenite or amber though as they will dissolve) in purified water, salt water or holy water. Smudging (using the smoke from white sage, dried herbs or incense) can also cleanse your crystals. Some people even give their healing crystals a moon bath by letting them sit out at night under the light of the full moon.
>
> After you've cleaned your stones and crystals for healing, align them with your energy by holding them in the palm of your hand, closing your eyes, stating your intention for them and thanking them for the healing they will provide.
>
> Finally, clean your crystals after each use. This means crystals you wear every day should be cleaned before you wear them the next day, and crystals used for a healing session should be promptly cleaned after each session.[2]

This entire paragraph is complete gibberish, a mishmash of scientific ideas blended with pseudoscience and mumbo jumbo. First of all, a crystal is just a regular arrangement of atoms at the molecular level. It has no power to absorb energy or release energy or do any of the other remarkable things claimed in these crystal healing sites. A "moon bath" is even more laughable, since the amount of energy coming from the reflected sunlight of the moon is minuscule and can barely be measured except by the most sensitive instruments.

Unless the crystal is highly porous (there are a few examples, such as clay minerals or rocks like sandstones and pumice), it does not absorb anything, not even water. Most of the hard stones used in the crystal healing sites are quartz, tourmaline, and others that have no porosity and can absorb nothing. Thus, cleaning them will also do nothing more than remove your sweat and body oils that built up as you've fondled them. Nothing washes any energy out of the crystal, since it was never absorbing human energy in the first place. Finally, it's amusing that they have to remind these gullible people not to wash certain stones (like selenite, a form of gypsum, or amber), since indeed these substances are water soluble and will magically vanish if you soak them in water.

Many of these sites make pseudoscientific lists of the "magical properties" of different stones, with fanciful descriptions of their powers, such as the following:

Listed below are my top seven crystals that can lead you to happier life.

1. Clear Quartz

Clear Quartz is known as "The Master Healer." It can be used for anything and everything. Clear Quartz is translucent and clear, and can heal issues at the physical, mental, emotional or soul level. It resonates with the higher chakras, bringing in divine white light and connection to higher-self, higher consciousness, higher wisdom and unconditional pure love. Clear Quartz can be programmed by a healer for just about any issue from your past.

2. Amethyst

Amethyst is known as "The All Purpose Stone." This stone is available in various hues and shades of light violet color, to lilac to vibrant purple. It resonates with Crown Chakra (Sahasrara), as well as Third Eye Chakra (Ajna), which opens up the gateway to divine consciousness and higher intuition. It also provides clarity when there's confusion in the mind, and helps to relieve stress and anxiety. Amethyst can even help with cell regeneration, insomnia, mood swings, and immunity. It is also known as the "Traveler's Stone," providing extra protection while you're out on the road, exploring new places.

3. Rose Quartz

This pink crystal is a very soothing and calming stone, symbolizing love and harmony. It helps open your heart to give and receive love. It also encourages you to forgive others and especially yourself, helping you to move on. It emits vibrations of love, harmony and peace.

4. Citrine

This golden, yellowish crystal is among the most beloved of all crystals. Everyone loves Citrine for its beauty and its healing properties; especially for those who want to invite more prosperity and abundance into their lives. Citrine is known as the "Merchant's Stone," as it can help you manifest more money by removing financial blockages, as well attract new opportunities. It is also considered one of the best stone for protection and weight loss. For abundance and prosperity, simply wear it or keep it in your purse or wallet, or place it in a cash drawer or money box in a corner of your home or office. Be as creative as you want.

For aura protection, wear it or keep in your purse or pocket. You can also place it at the four corners of your house, or in a place where there are a lot of activities. As a weight-loss stone, place it on your Solar Plexus Chakra (Manipura) daily for about 20 minutes, while you visualize your goals.

5. Black Tourmaline

This stone is definitely one of my personal favorites—I have one placed in every room of my house! My children and I always wear it to keep it close to our aura for protection. Black Tourmaline is considered among the top protective stones. It can also help you with grounding and create shield against harmful electro magnetic fields.

Place this stone near your electronic devices and gadgets to protect you from EMF frequencies. Place it in your house or workplace to protect you from the negative vibrations of others, or any form of potentially harmful intentions that might be directed toward you. Wear it to absorb negativity and transmute it to positivity energy. If you have hard time manifesting your wishes, write your intentions on a small piece of paper and put it underneath Black Tourmaline. This will help remove any energetic or psychic blockages related to your wish, filling it up with positive energy and abundance.

6. Carnelian

This stone color varies from a light orange to reddish brown. Keeping this stone around you keeps you motivated, inspired and confident. It will give you the perseverance you need to not give up on any tough situation. Carnelian is an energy booster that spreads joy by burning away stuck and impure energies.

This stone represents the Sacral Chakra (Svadhisthana) which can help alleviate addiction, and harness your sensuality and creativity.

7. Aventurine

Aventurine comes in various colors such as green, red, yellow, peach, blue and more. Aventurine attracts true love, true friendship and lasting relationships. Green Aventurine is called the "Luck Stone," bringing forth good fortune and new opportunities.

It also increases your confidence and self-esteem, optimizing your personal growth. Aventurine can be used for heart-related ailments and emotional issues, as it represents the color of Heart Chakra (Anahata). Use this stone to help with your allergies, sleep disorders, immunity and to regulate your blood pressure.

There are innumerable crystals with an abundance of healing properties, each with different colors, hues and shades. Choose the ones that resonate the most with you and use them along your healing journey toward a happier, healthier life![3]

This entire piece is pure pseudoscientific garbage from beginning to end. First, there really isn't much difference between quartz crystals of different colors. All of the mineral varieties of quartz (including rose quartz, gray smoky quartz, purple amethyst, yellow citrine quartz, orange carnelian quartz, and green aventurine quartz) all have exactly the same lattice of silicon dioxide typical of quartz. The only minor differences (less than 1% of the entire crystal) are small amounts of impurities, like iron or manganese or bubbles of water, which give the quartz crystals their different colors. For example, aventurine has tiny flakes of mica in it that give it its shimmer, while carnelian is red and citrine is yellow due to iron oxides (better known as rust). (Natural citrine is quite rare, actually, so there is a huge industry of baking amethysts in an oven until they turn yellow and become citrine, all to feed the market of healing crystals.) Thus, at the atomic level, there is no reason to think that clear quartz would have any significantly different properties than purple amethyst quartz, yellow citrine quartz, orange carnelian, or green aventurine.

What Does Science Say?

The world of healing crystals operates completely outside of legitimate science and medicine. This is apparent from the site about the seven best healing crystals, which tacitly admits at the end, "This post is not intended for use as medical advice."[4] Many internet sites say in fine print at the bottom that they are for entertainment purposes only and should not be considered a substitute for regular medical treatment, even if the rest of the site is acting as a replacement for real medicine. Indeed, they have to put a warning like that somewhere, because they are financially and medically liable if they make specific promises and medical claims and if someone is harmed and sues them for damages.

A neutral site, Answers.com, points out that "medical professionals place little credence in crystal therapy, attributing any observed benefits to placebo effect. Their skepticism stems from a lack of scientific evidence for the healing effects of crystals, and from differences of opinion among practitioners about how the therapy actually works."[5] One of the crystal sites, IndianReikiMasters.com, admits "the main problem in providing hard core scientific data that would be acceptable to scientists at large is that the sensitive equipment necessary to measure these energy changes is not available today. . . . If the energy could not be measured, the effect it had on the human body could be. Therefore, this inductive means of study became our laboratory. Unconventional devices such as dowsing rods were used to measure the expansion or contraction of the body's energy field."[6] This is a tacit admission that nothing they claim can be tested by scientific methods, so it is completely outside the realm of real, testable science-based medicine and instead lurks in the murky world of pseudoscience.

So how is it that people who treat themselves with crystal healing often feel better? The answer is a classic example of how quack medicine fools people into believing that their treatments are real: a combination of the placebo effect (giving a patient something harmless and ineffective and using the power of suggestion to somehow help healing), plus confirmation bias (noticing the good results and ignoring the problems or missed predictions), plus wishful thinking. An article in *Live Science* puts it this way:

While there are no scientific studies on the efficacy of crystal healing, there is a study that suggests that crystal healing may induce a placebo effect in a patient who receives this type of treatment. Placebo effects are effects that accompany a treatment that are not directly due to the treatment itself acting on the disease of the patient, according to Christopher French, head of the anomalistic psychology research unit at the University of London. In other words, a person may feel better after undergoing crystal healing treatment, but there is no scientific proof that this result has anything to do with the crystals being used during the treatment. In 2001, French and his colleagues at Goldsmiths College at the University of London presented a paper at the British Psychological Society Centenary Annual Conference in Glasgow, in which they outlined their study of the efficacy of crystal healing. For the study, 80 participants were asked to meditate for five minutes while holding either a real quartz crystal or a fake crystal that they believed was real. Before meditating, half of the participants were primed to notice any effects that the crystals might have on them, like tingling in the body or warmth in the hand holding the crystal.

After meditating, participants answered questions about whether they felt any effects from the crystal healing session. The researchers found that the effects reported by those who held fake crystals while meditating were no different than the effects reported by those who held real crystals during the study. Many participants in both groups reported feeling a warm sensation in the hand holding the crystal or fake crystal, as well as an increased feeling of overall well being. Those who had been primed to feel these effects reported stronger effects than those who had not been primed. However, the strength of these effects did not correlate with whether the person in question was holding a real crystal or a fake one. Those who believed in the power of crystals (as measured by a questionnaire) were twice as likely as non-believers to report feeling effects from the crystal.

"There is no evidence that crystal healing works over and above a placebo effect," French told Live Science. "That is the appropriate standard to judge any form of treatment. But whether or not you judge crystal healing, or any other form of [complementary and alternative medicine], to be totally worthless depends upon your attitude to placebo effects." As French pointed out, there are many forms of treatment that are known to have no therapeutic effect other than a placebo effect. However, while these treatments might make you feel better temporarily, there is no proof that they can actually cure diseases or treat health conditions. If you're suffering from a serious medical issue, you should seek treatment from a licensed physician, not an alternative healer, French said.[7]

To sum up, crystals are pretty and fun to own, but don't fall for the bogus claim that they have magical powers or use them instead of real medicine to treat real illnesses. Don't be like the famous comedian Andy Kaufman, who rejected real medical treatment and used crystal healing—and died because he wasn't taking his real medicine anymore.

This site puts it all in perspective:

A fact unknown to crystal advocates is that every mineral grain, no matter how small, retains its crystal lattice, and so should retain all the magical powers of that mineral. Given that the entire land surface of the Earth is made of rocks, sediments and soils that are largely mineral grains, everyone on the planet should be receiving the benefits without having to purchase overpriced and poor quality crystals from scented shops or people with purple hair. Beaches are often almost entirely quartz grains, as are deserts, and so overall there should be no negative energy at all anywhere on Earth.[8]

Water Witching

Forked Sticks and Fakery

One of the oddest practices that actually attempt to discover something real about the earth (rather than speculate blindly based on the Bible or conspiracy thinking) is dowsing, also known as divining, water witching, doodle-bugging, and a variety of local names. In a nutshell, the practitioner walks around some kind of divining rod (*virgula divina* in Latin) such as a forked stick or pair of L-shaped rods clutched in his or her hands (fig. 17.1). When the rod suddenly points down, the dowser will claim that something of value is in that spot. People usually use dowsing to find places with underground water to dig a well, but it has also been used to find veins of precious gems or metals or to find oil deposits.

As written in Samuel Sheppard's 1651 publication, *Epigrams Theological, Philosophical, and Romantick*:

> Virgula divina.
> Some Sorcerers do boast they have a Rod,
> Gather'd with Vowes and Sacrifice,
> And (borne about) will strangely nod
> To hidden Treasure where it lies;
> Mankind is (sure) that Rod divine,
> For to the Wealthiest (ever) they incline.[1]

The practice of divining to find metal deposits can be traced back to the 1500s in Germany, and in the same century, German diviners were licensed in England to find deposits of tin in Cornwall, lead in the Mendip Hills, and silver in Wales. One of the oldest known books in the history of geology, Georgius Agricola's *De Re Metallica* (1556), described how dowsing was used to find metal ore:

Figure 17.1. Illustration from Agricola's 1556 *De Re Metallica*, showing dowsing as it was practiced at the time. (*Courtesy Wikimedia Commons.*)

There are many great contentions between miners concerning the forked twig, for some say that it is of the greatest use in discovering veins, and others deny it. . . . All alike grasp the forks of the twig with their hands, clenching their fists, it being necessary that the clenched fingers should be held toward the sky in order that the twig should be raised at that end where the two branches meet. Then they wander hither and thither at random through mountainous regions. It is said that the moment they place their feet on a vein the twig immediately turns and twists, and so by its action discloses the vein; when they move their feet again and go away from that spot the twig becomes once more immobile. . . . [Those seeking minerals] should not make use of an enchanted twig, because if he is prudent and skilled in the natural signs, he understands that the forked stick is of no use to him.[2]

Even though Agricola was one of the first people to publish a book on aspects of geology and many of his ideas were largely wrong and based on a medieval concept of the earth, rocks, minerals, and fossils, he was correct about "the enchanted twig." Divining was equally outdated. The famous scholar Paracelsus said that divining rods had "deceived many miners" and that "faith turns the rod."[3] The most famous translation of Agricola was done by Herbert Hoover, then a mining engineer who eventually was elected president of the United States in 1928, just before the Great Depression. Hoover pointed out in a footnote to his translation: "There were few indeed, down to the 19th century, who did not believe implicitly in the effectiveness of this instrument, and while science has long since abandoned it, not a year passes but some new manifestation of its hold on the popular mind breaks out."[4]

Agricola and later authors tried to explain how divining for minerals worked by suggesting that minerals produced mysterious emanations that were detected by the rod. According to William Pryce, who wrote about it in 1778 in *Mineralogia Cornubiensis*,

The corpuscles . . . that rise from the Minerals, entering the rod, determine it to bow down, in order to render it parallel to the vertical lines which the effluvia describe in their rise. In effect the Mineral particles seem to be emitted from the earth; now the Virgula [rod], being of a light porous wood, gives an easy passage to these particles, which are also very fine and subtle; the effluvia then driven forwards by those that follow them, and pressed at the same time by the atmosphere incumbent on them, are forced to enter the little interstices between the fibres of the wood, and by that effort they oblige it to incline, or dip down perpendicularly, to become parallel with the little columns which those vapours form in their rise.[5]

But in 1518, Martin Luther condemned dowsing as an occult, Satanic practice that broke the first commandment. In 1662, the Jesuit priest Gaspar Scott condemned it as Satanic, yet in the 1600s divining rods were also used to determine the guilt or innocence of criminals and heretics on trial in southern France. The practice was so badly abused that by 1701, the

Inquisition forbade it to be used to determine criminals and seek justice. Remember, when this was happening, it was still common to use practices like dunking women in a pond or well to determine whether they were witches, so dowsing was no more bizarre than other ideas of that time. Since then, divining rods of various kinds have been used to find lost objects, hidden drugs, diseases in people, and even hidden explosives.

Water, Water, Everywhere

About the same time as these beliefs were common, divining began to be used to find water also. The oldest-known written account of this practice dates to 1568. According to Sir William F. Barrett's 1911 book *Psychical Research*,

> In a recent admirable *Life of St. Teresa of Spain*, the following incident is narrated: Teresa in 1568 was offered the site for a convent to which there was only one objection, there was no water supply; happily, a Friar Antonio came up with a twig in his hand, stopped at a certain spot and appeared to be making the sign of the cross; but Teresa says, "Really I cannot be sure if it were the sign he made, at any rate he made some movement with the twig and then he said, 'Dig just here'; they dug, and lo! a plentiful fount of water gushed forth, excellent for 'drinking, copious for washing, and it never ran dry.'" As the writer of this *Life* remarks: "Teresa, not having heard of dowsing, has no explanation for this event," and regarded it as a miracle. This, I believe, is the first historical reference to dowsing for water.[6]

Dowsing for water is by far the most common use of divining rods in the past century. Despite being debunked many times, they are still widely used by people who don't trust experts but only the folk wisdom they learned from their community. Dowsers used the traditional forked sticks of the early diviners looking for mineral veins or, in later years, they held two L-shaped rods (fig. 17.2A, B) with the short ends in each hand and the long ends supposedly swung around until they pointed to water.

So if the rods really don't work, how is it that they sometimes do find water? It turns out that these dowsers have an entirely false notion of where water is found underground. Because the practice came from finding veins of minerals, they think that water underground is found in veins and underground rivers, flowing away from domes of water. But this is completely wrong, as hydrologists have known for over a century.

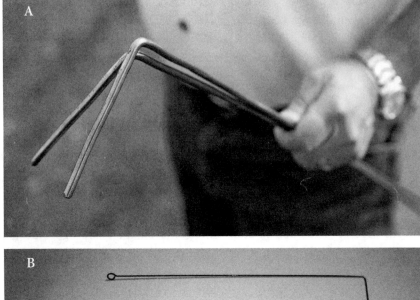

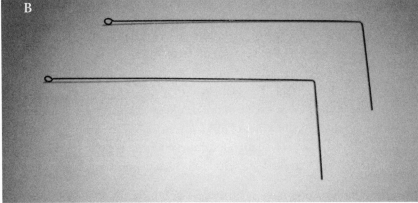

Figure 17.2. A, B. A pair of L-shaped metal divining rods. The short segment is held in each hand, and then the long ends supposedly swing and point toward water. (*Courtesy Wikimedia Commons.*)

As rainfall seeps into the ground, it flows down (infiltrates) through the pore spaces between the sedimentary grains in the subsurface (fig. 17.3). This unsaturated soil and sand with mostly air in the pore spaces is known as the zone of aeration. Eventually water accumulates deeper underground to form a zone of saturation and fills all the pore spaces. The top surface of the saturated zone is called the water table, which is a dynamic feature, moving up and down in response to rate of inflow of water. If there are heavy

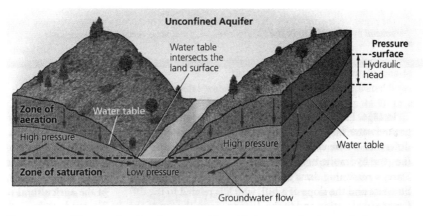

Figure 17.3. Diagram showing the water table, with the water flowing to the lowest spot on the water table. (*Courtesy Wikimedia Commons.*)

rains, the water table will rise, and sometimes it will completely saturate the highest layers of soil and flood the entire landscape. During dry periods, the water table will drop.

In addition, the water table is generally not a flat surface but forms underground "hills of water," or high spots, and "valleys," or low spots. The hills of water roughly correspond to the topographic hills and ridges, where water flowing down from the tops of rainy hills must sink down from a higher elevation to reach the water table. Consequently, in any zone of saturation, gravity drives a natural flow of water from the high spots in the water table down to the low spots (fig. 17.3). Low spots often correspond to the rivers and lakes on the landscape, which are replenished by groundwater flowing down to them even when there is no rain for a long time. Where groundwater appears at the surface, it forms a spring.

Not all kinds of subsurface material transmit water equally well. A very porous and permeable material, like the sand and gravel generally underlying most valleys or a porous sandstone or rock shattered with lots of cracks, is an aquifer (Latin for "water bearing"). Sediments or rocks that are nonporous and impermeable, so water cannot pass through them, are known as aquicludes (Latin, "closed to water"). Layers of mud or shale are the most common types of aquicludes, along with solid rock masses with no pores or cracks. A rock or sediment that lets water slowly pass through is called an aquitard.

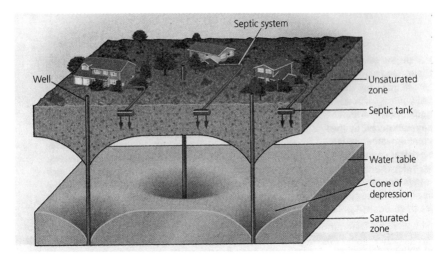

Figure 17.4. Diagram showing the effect of pumping a well to create a cone of depression. A very deep well that is strongly overpumped will depress the water table and pull all the flow from shallower wells to the deepest well. (*Courtesy Wikimedia Commons.*)

If people dig or drill wells down to the water table and pump it faster than it is replenished, the water table will drop (fig. 17.4). Lots of heavy pumping in the well will cause the place where the well bottom penetrates the water table to suck all of the water around it and form a depleted area at the well called a cone of depression. This is a common problem where water is scarce. The first wells in an area may not be that deep and may just barely reach the water table. If they pump water too fast in comparison with its rate of replenishment, or if newer deeper wells are installed that pump even faster, the water table will drop and the shallow wells will go dry (fig. 17.4).

Water always flows downhill or toward the direction of least pressure. Thus, if you sink a very deep well and pump very hard, the water will flow away from the higher parts of the water table and the shallower wells and down to the deepest well that depresses the water table the most. This can create problems. If, for example, the natural flow of water goes away from a source of contamination and then the water pumping changes the high spots on the water table, the flow will reverse and contaminants will be sucked into the deeper wells (fig. 17.4).

Rather than rivers or veins of water underground (as traditional dowsers long conceived it), it is more accurate to think of underground water as a giant flat sheet of saturated sand and soil beneath your feet in just about any place where you can stand on loose soil (rather than hard impermeable bedrock). There is water beneath you nearly everywhere, so when a dowser's stick points down, you are guaranteed to find some water if you're willing to dig or drill deep enough. Thus, you don't need to waste money paying a fake like a dowser to find water when it is almost everywhere beneath your feet.

However, there are ways, using genuine hydrological methods, to find how deep the water table is and how far you would need to drill to reach it. Genuine scientists who study how groundwater really moves, called hydrogeologists, use a variety of real methods, such as plotting the water table between wells, testing the chemistry of the well to determine flow and contamination, and utilizing real underground sensing methods to locate where the water table is most accessible and to determine whether pumping might cause problems of contamination.

Thus, dowsers are pretty much guaranteed to find some water if they dig deep enough, but if they don't find it right away, they have ready-made excuses and ad hoc rationalizations ready to explain their apparent failure. As geologist and skeptic Sharon Hill wrote,

> Judging a successful well isn't so clear cut and this allows dowsers some wiggle room with their claims. If it's not immediately clear that a good yield has been hit, give it a few days or use well development techniques and it will probably turn out OK. Or maybe, you just need to go a bit deeper. Dowsers might also claim that they correctly spotted a vein but it was crushed or diverted by the drilling. In any case, the failures are overshadowed by bragging about successes.
>
> Some dowsers attempt to uses sciencey-sounding explanations for their practice. Dowsers variously claim that they are influenced by "electricity" or "magnetism" or "chemistry," that they feel "energy" over the spot. (It may be called "radiesthesia" or "cryptesthsia.") None of these signals are given by water in a degree sufficient for human detection. Instrumentation is more sensitive to small environmental changes. Yet, instruments can't objectively detect these dowsing "energies" to match the claims. Case histories and field tests provide the bulk of positive evidence for dowsing. Such results are subjective, flawed, and not scientific. A baseline for comparison is missing. Consistency is not taken into consideration. Authors Vogt & Hyman think that bias is clearly in play when dowsers take credit

for finds—observers are impressed by dramatic and rare events and forget the failures. They also ignore the ease of finding water in many areas where the feat is really nothing special. Dowsers consistently and reliably fail when asked to find water in blinded tests. More damning is that in objective tests, two or more dowsers will not converge on the same prime location![7]

So the presence of groundwater beneath you nearly everywhere explains part of the "success" of dowsing or water witching. But what about the strange movements of the divining rod when it suddenly points down? Research suggests it may be a result of involuntary muscular motions known as the *ideomotor effect*, where small muscle twitches in the arms and hands (sometimes not even in the conscious control of the dowser) cause the stick to twitch up and down. It's analogous to the way the pointer on the board of the Ouija game board works. If the people whose fingers are pressed on it are not consciously pushing it to form a message, nonetheless the involuntary ideomotor motions of their hands and arms will cause it to twitch and slide.

As Sharon Hill wrote,

> Dowsers claim they have a gift or inherited power to find water. The rod, they say, amplifies this power to detect the material which makes it move so dramatically. The movement, however, is more naturally explained. The forked stick is held in such a way, palms up, with tension in the muscles, that any tiny movement in the wrist or arm muscles will cause it to unwind and react, like a coiled spring. The stresses of this unnatural position are released without the person even being aware they are doing it. The same holds true for metal rods balanced in loose grips that allow free movement. The smallest movement will cause a reaction in the rods. This involuntary motor behavior is the same effect that causes a pendulum or Ouija board planchette to move. Implicit muscle activity occurs just by THINKING of movement.[8]

How Do We Know?

So how do we know that dowsing is really pseudoscience? The best way is to put it to the test and to see if dowsers can find something with a success rate greater than chance alone could explain. Such tests have been done many times, and in every case, dowsing flunked the test:

- Almost a century ago in 1927, Smithsonian geologist John Walter Gregory found that dowsing was simply a matter of chance, often helped by using surface clues (like nearby springs).[9]

- During World War II in 1943–1944, geologist W. A. MacFadyen tested three dowsers working in Algeria, just after it had been liberated from the Germans. All of their results were negative.[10]

- In 1959, British dowser P. A. Raine was tested and failed to find the location of a buried kiln full of water that was successfully located by a magnetometer.[11]

- In 1971, the British Ministry of Defence commissioned engineer R. A. Foulkes to test numerous dowsers. Once again, the results were "no more reliable than a series of guesses."[12]

- Physicists John Taylor and Eduardo Balanovski performed a series of experiments in 1978 looking for unusual electromagnetic fields from dowsing subjects, but they did not detect any.[13]

- Evon Vogt and Ray Hyman, in the study mentioned by Sharon Hill, conducted a 1978 survey of many controlled studies of dowsing for water and found that none of them gave results better than chance.[14]

- A study by Richard N. Bailey, Eric Cambridge, and H. Denis Briggs on dowsing on the grounds of various churches claimed successful results, published in their 1988 book *Dowsing and Church Archeology*. Archeologist Martijn Van Leusen reexamined their work and concluded that the studies were badly designed, with the authors unfairly redefining their test parameters on hits and misses to give positive results.[15]

- In 2006, a report examining fourteen published studies of grave dowsing in Iowa found none of them located human burials correctly, and simple scientific analysis showed that their "principles" for explaining grave dowsing were nonsense.[16]

- In a famous 1991 study, magician and skeptic James Randi (better known as the "Amazing Randi") offered a $10,000 prize to any successful dowser. The study, conducted in Kassel, Germany, involved some thirty dowsers who had to locate water in buried plastic pipes; the experimenters could route the water through different pipes, so the dowser could not know which ones were empty and which ones were full. Before the experiment, all the dowsers signed a statement agreeing that this would be a fair test of their abilities and that they expected to be successful 100 percent of the time. After all the experiments were over, none of the "successes" were better than chance.[17]

- An even more complex study was run in 1987–1988 in Munich, Germany, by Hans-Dieter Beck. For this experiment, some five hundred dowsers participated; then after eliminating most of them, forty-three were selected as more skilled than the rest. The test was conducted in a large barn with a series of pipes that could be filled or emptied and moved as well. Most of the dowsers (thirty-seven out of forty-three) showed no results better than chance, but Beck insisted that six of them were actually able to perform better than chance would dictate.[18] However, physicist Jim Enright reexamined the entire study and showed that it was flawed. The "successes" of the six "genuine dowsers" could be explained by random fluctuations in large data sets, and none of the tests could be explained by anything beyond random chance.[19]

Divining Never Dies

Despite the thorough debunking that has been performed again and again about dowsing for water and using divining rods to find objects, the idea keeps popping up again and again—often in strange places. No matter how much we try to test it and prove it wrong, the power of wishful thinking, confirmation bias, and irrational nonmathematical thinking allows people to be suckers for scams of objects that are supposedly able to detect things we cannot sense. Dowsers are still working actively in places all around the world, especially where local farmers distrust experts and are more influenced by traditional beliefs and practices than they would be by the arguments of a skeptic trying to convince them they are being defrauded.

The latest fad is for companies to sell detection devices to police departments, military units, and security companies to find things like explosives. In the rush to fight terrorism and deal with bomb disposal, a lot of legitimate organizations have been fooled and wasted tax dollars on worthless fake detectors that were no more than expensive frauds. Some examples include the following:

- The MOLE Programmable System built by Global Technical Ltd. of Kent, England. It was tested by Sandia National Laboratories and found to be worthless.[20]

- A device called the ADE 651 manufactured by ATSC of the United Kingdom and sold in large numbers to the Iraqi police to detect explosives. Despite its widespread adoption, it failed to prevent many bombings in Iraq, costing lots of people their lives. The device was tested by several groups and determined to be ineffective, and the British government banned its export. Eventually, the director of ATSC was convicted of fraud and sentenced to ten years in prison.[21]

- Another contraption, called the Sniffex, was used to sniff out bombs but turned out to be worthless. The US Navy Explosive Ordinance Disposal unit concluded that "the handheld Sniffex explosives detector does not work."[22]

- Yet another device, the Global Technical GT 200, was a virtual dowsing rod that claimed to detect explosives. When studied more closely, it was found to have no actual scientific mechanism to make it work; it was no better than a forked twig.[23]

In this age of terrorism and bomb scares, there is always a market for scam artists attempting to make a buck off frightened government organizations and their officers who must deal with bomb disposal and detection. The situation has gotten so bad recently that the US government had to publish a study warning police departments and security organizations about such frauds.[24]

Dowsing may go back to the Middle Ages, but the folk beliefs in its powers never seem to die, no matter how it is tested. As we have seen with the fake "security" devices, there is no shortage of con artists who will claim to have devices that can magically detect things—but are actually useless.

Mysterious Earth
Why People Believe Weird Things

Why Do People Believe Weird Things?

When most people hear that there are groups of educated Americans or Europeans in the twenty-first century who still believe in things like the flat earth, or the geocentric universe, or the idea that the earth is only six thousand years old, they shake their heads in disbelief. They cannot imagine how anyone with any kind of modern education would take such things seriously. When I mention it to other people, I'm often asked, "What are they thinking? How can they take such ancient outdated notions seriously in the age of high technology and instant information at our fingertips? How could anyone believe such things?"

It seems even more ironic that superstitions like Young-Earth Creationism or prescientific notions like the flat earth and geocentrism have any followers at all in this age of nearly universal education and booming technology that spreads ideas around the world, allowing us to access information in seconds via the internet that used to take from hours to years to find. In fact, at the dawn of the scientific revolution and the Age of Enlightenment, many scholars and scientists were optimistic that reason and science would triumph over superstition, prejudices, and irrationality. During the late 1700s, the famous French philosopher, mathematician, and social theorist the Marquis de Condorcet wrote in his 1795 work *Sketch for a Historical Picture of the Progress of the Human Spirit,*

> Our hopes for the future condition of the human race can be subsumed under three important heads: the abolition of inequality between nations, the progress of equality within each nation, and the true perfection of mankind. Will all nationals one day attain that state of civilization which the most enlightened, the freest and the least burdened by prejudices, such as the French and the Anglo-Americans, have attained already? Will the vast gulf that separates these peoples from the slavery of nations under the rule

of monarchs, from the barbarism of African tribes, from the ignorance of savages, little by little disappear?

The time will therefore come when the sun will shine only on free men who know no other master but their reason; when tyrants and slaves, priests and their stupid or hypocritical instruments will exist only in work of history and on the stage; and when we shall think of them only to pity their victims and their dupes; to maintain ourselves in a state of vigilance by thinking on their excesses; and to learn how to recognize and so to destroy, by force of reason, the first seeds of tyranny and superstition, should they ever dare to reappear amongst us.[1]

And in another passage, Condorcet wrote,

As the mind learns to understand more complicated combinations of ideas, simpler formulae soon reduce their complexity; so truths that were discovered only by great effort, that could at first only be understood by men capable of profound thought, are soon developed and proved by methods that are not beyond the reach of common intelligence. The strength and the limits of man's intelligence may remain unaltered; and yet the instruments that he uses will increase and improve, the language that fixes and determines his ideas will acquire greater breadth and precision and, unlike mechanics where an increase of force means a decrease of speed, the methods that lead genius to the discovery of truth increase at once the force and the speed of its operations.[2]

Sadly, Condorcet's optimism about the spread of rationality among free, educated humans and their societies was discredited by the French Revolution and the Reign of Terror. Condorcet was initially a supporter of the revolution and the new birth of freedom from the old order it represented; however, in 1793, the political winds turned against Condorcet when he opposed the assembly's vote to execute King Louis XVI. Condorcet was forced to flee to avoid arrest. His 1795 *Sketch* was written while he was in hiding and published posthumously because he eventually was caught and died in prison in 1794.

Like the Marquis de Condorcet, many later, more optimistic philosophers and social theorists held on to the view that once humanity had reached the point where education was universal and science was valued as one of our greatest gifts, then the Age of Enlightenment, which had broken out of the shackles of religious and political tyranny in the 1700s, would reach its completion. Yet here we are in the twenty-first century, and irrational behavior and primitive superstitions are still rampant among us.

So why do people reject science and reality and believe weird things? As years of research have shown, the answer is complicated. The primary factor is that humans are not as rational as we would like to think. As Michael Shermer argues in his book *The Believing Brain*, we are "belief engines": our brains operate using a set of core beliefs and assumptions that we hold without question, what is often called our "worldview," and then we filter what we see around us to fit what we already accept as true.[3] This is what psychologists call motivated reasoning: we are not purely rational beings; we accept or reject ideas and information motivated by our existing biases (often unconsciously).

Consequently, one of the classic Enlightenment views—that rational arguments and evidence will eventually win out—turns out to be wrong in many cases. For people who have strong emotional and community connections to a belief system (whether it be a religion or a political party or whatever), their minds are preparing arguments against anything that weakens or challenges that belief (like a lawyer preparing his slanted case to defend one side of an argument), not listening to reason or evidence. This is really discouraging to those of us who are battling irrationality. According to the Enlightenment view, truth, reason, and evidence should eventually persuade anyone and win out, but what psychologists have shown is that die-hard creationists, climate deniers, flat-earthers, and geocentrists (along with other true believers) cannot be persuaded this way.

The most obvious symptom of motivated reasoning is what psychologists call *confirmation bias*: we remember what confirms or agrees with what we believe, and we forget, deny, or don't even notice things that conflict with our existing beliefs. Confirmation bias is how psychics and fortune-tellers fool their marks. If you watch the unedited video of one of these con artists in action, the psychic will start with lots of guesses based on vague generalities that are true of most people. If psychics miss, they quickly shift to something else, until they get a hit in the form of a positive response from their mark. They then follow the clues the victim gives them, quickly shifting to whatever gets a positive response or favorable body language, until they can narrow it down. The mark is amazed that the psychic knew so much about them, but once you watch the unedited video, you realize that it was all guesswork based on things that are generally true of most people, followed up by clues that the victim provides, often unconsciously or unintentionally. (For that reason, when psychics are shown on TV, they often edit out the misses and focus on the apparent hits.)

Likewise, when we hear something on the news that agrees with our worldview, we pay attention and may even absorb it and use it ourselves. When we run into something that conflicts with our beliefs, we tend to ignore it, dismiss it, or (rarely) try to dispute it. In our modern world, with polarized media where we can hear right-wing viewpoints on Fox News and other viewpoints on other networks, we consciously set our channels to give us information that is already filtered to fit our existing biases. A good example, as Matt Taibbi showed in his book *The Great Derangement: A Terrifying True Story of War, Politics, and Religion*,[4] is the Young-Earth Creationists. They often live in highly self-isolated church communities where they avoid any media or books or anything that might challenge their views, and they only read or listen to what their church tells them to. They spend most of their free time at meetings with fellow church members, so they get constant reinforcement of their beliefs and never hear anything that challenges their ideas—and their pastors often forbid them to even listen to anything outside their church.

Rejection of false ideas and confirmation bias doesn't necessarily correlate with intelligence. People who are bright but also deeply committed to a bizarre belief system can create incredible justifications and rationalizations as to why their preexisting beliefs are right and their opponents' beliefs are wrong. As Shermer put it, "Smart people believe weird things because they are skilled at defending beliefs they arrived at for non-smart reasons."[5] Thus we have "smart idiots"—people who are actively engaged in an argument, well educated, and smart by any standard measure but who selectively bias what they learn so that they can argue against reality in defense of their community and belief system. I've debated clever creationists who can make ridiculous arguments sound plausible to the listener who doesn't know the facts about the situation.

Often, confronting dogmatists and telling them that their information is wrong creates a backfire effect; they just become more entrenched in what they believe and even cleverer about rationalizing and justifying it.[6]

It's usually pointless to argue facts with someone who is deeply committed to demonstrably false beliefs, because you will never convince that person. You'll just waste your time, and in many cases, you will drive these believers away from even considering that they might be wrong.

The second major example of motivated reasoning to preserve core beliefs is called *reduction of cognitive dissonance*. Despite the fact that we like to think of ourselves as consistent in our beliefs and behavior, psychologists have found that the brain is very good at compartmentalizing all the different ideas and beliefs we have. Sometimes they conflict with one another, creating an uncomfortable feeling, and then we must do what we can to reduce this cognitive dissonance. For example, we all like to believe we are moral and honest and follow the rules, yet most of us break the rules once in a while in small ways, such as speeding when we think it's safe, or telling white lies to protect people's feelings and keep social situations tolerable. Studies show that people often cheat, even in small ways that seem harmless, when there are no harsh consequences. One part of our brain is telling us that we are good people and don't cheat or lie or break the law, but in another part of our brain we remember doing it. To reduce this dissonance between two beliefs or ideas, we find a way to rationalize or justify our behavior and thus reduce the clash or dissonance in our brains.

There are lots of good examples of this behavior. For example, every smoker has heard again and again the warnings that smoking is bad for you. Smokers then either decide to quit smoking or (thanks to the addictive powers of nicotine) they rationalize their smoking by saying, "The studies might be wrong or inconclusive." The climate denier is usually deeply committed to a belief in unfettered free-market capitalism and a fear of any kind of government intervention, even if it means reducing pollution or saving the planet for future generations. Climate deniers go through elaborate false rationalizations to discredit the consensus of the world's climate scientists by claiming, "The science isn't settled" or "Scientists are getting big grants to destroy capitalism" or "It snowed yesterday; therefore, global warming is a hoax."

We have seen this in many of the examples we have discussed in this book. For example, creationists believe that a literal interpretation of Genesis is essential to their faith and to their psychological well-being. Anything that challenges their faith therefore must be wrong or the work of the devil. If you believe you are going to hell for accepting the scientific facts of evolution, would you be swayed by facts or arguments, or would you grasp at any straws to deny scientific reality and save your soul?

This has been documented many times. For example, British science journalist Bruno Maddox visited the creationists of Ken Ham's Answers in Genesis organization and had this account:

> I find myself reminded of F. Scott Fitzgerald's proposition in *The Crack-Up*, that "the test of a first-rate intelligence is the ability to hold two opposed ideas in mind at the same time and still retain the ability to function." Fitzgerald's first-rate mind, of course, eventually stopped retaining its ability to function, and watching [creationist Jason] Lisle try to reconcile the cutting edge of modern planetary physics with the offhand assertions of a religious tract written thousands of years ago by an unknown assortment of bearded semi-cave dwellers, I found myself wondering how long the poor chap has.
>
> For the record, I have even less patience now with the creationist agenda than I did going in, because I now suspect that they don't really believe the falsehoods with which they are trying to flood the world. But at the same time I got the clear impression that they don't have any choice. I thought I was going to meet people who love God and therefore hate science. What I found instead were people who love God but who have at least a pretty serious crush on science as well, and thus find themselves in the Fitzgeraldian nightmare of waking up every day and trying to believe in both. They will—they must—spend their lives, and brains, trying to think of ways that patently false ideas can be made to seem, if not actually true, at least not quite so patently false. It is, I fear, a doomed exercise, but it's a heroic one as well, it pains me to admit.
>
> Not to overdo the Fitzgerald, but I shall think of [the creationists] often, as day after day they beat on, boats against the current of truth, borne back ceaselessly into being just completely, utterly wrong.[7]

Actually, Fitzgerald and Maddox are wrong: we *all* have the ability to hold two or even more incompatible ideas in our brains at the same time, and we do it constantly. But sometimes it is easier to reduce the clash between them than at other times.

Finally, the root of most of our cognitive biases is tribalism. We are all products of our families, our communities, and other groups to which we have belonged. We learn most of our worldview from them. We typically decide what is false and what is true based on what we were taught when we were young. We also view with suspicion ideas or beliefs of other groups who were raised differently from us or who come from different cultures. That is why what you believe is the "true religion" is largely based on where you were raised. If you grow up as a Christian in the Christian parts of the world, Christianity is "obviously true," and it is difficult to imagine anyone following other religious beliefs. But the same is true of someone growing up in a Muslim part of the world (where not being a devout Muslim can often be a death sentence), or growing up in the Hindu parts of India, or in any part of the world with one prevailing religion: all other religions other than that which you grew up with are clearly false. It takes lots of education in comparative religions plus an open mind (and often some world travel) to overcome this bias, and most devoutly religious people can't (or won't) even try to challenge the beliefs of their family, their friends, their neighbors, their town, and the church they grew up in.

Sources of Weird Ideas

These are the factors that help us understand why people believe weird things. Let's look at the spectrum of beliefs we encountered in this book and dissect their sources and motivations:

- Creationism: By far, the commonest motivation for weird ideas about the earth is Young-Earth Creationism. As we saw in the chapters on flat-earthers, geocentrists, flood geologists, young-earth beliefs, and the ideas that dinosaurs are faked, these are all inspired by a literal interpretation of the Bible. We have outlined the Bible verses that really do suggest that the earth is flat and the center of the universe, as well as plenty of verses describing Noah's flood, and we have examined the Ussher tradition that the earth is only six thousand years old. What is fascinating and ironic is that not all creationists agree on these ideas, even though they all claim to believe in the literal truth of the Bible. For example, most of the major Young-Earth Creationists today, like Ken Ham's Answers in Genesis organization, promote Noah's flood and the young earth, but they are adamant in trying to distance themselves from flat-earthers and geocentrists because they think such ideas might make them look foolish in the eyes of outsiders (as if believing in Noah's flood and a young earth don't make them look foolish enough).

- Occult/New Age/paranormal thinking: The second category of weird ideas seems to originate largely from people who believe in ancient legends and myths, sometimes updated with more recent ideas like UFOs and aliens. The hollow earth idea, along with the Atlantis and Lemuria myths, the aliens supposedly living in Mount Shasta, and much of the mythology associated with crystal healing all fall into this category. Ley lines started as a harmless false idea originated by amateur archeologists, but they are now fully absorbed into the paranormal culture. Often, these fantasies of paranormal and occult thinking overlap, so people who are inclined to believe in Atlantis and UFOs are also susceptible to believing in the magic of crystal healing.

- Conspiracy theorists: Many of the ideas discussed in this book have a component of conspiracy thinking in them. The moon-landing deniers are pure conspiracy theorists, and many of the other movements that are rejected by the mainstream blame their rejection on a grand conspiracy by scientists or all of society to suppress their ideas. This excuse is used by flat-earthers and geocentrists, who, like moon-landing deniers, also believe that everything from NASA is a hoax and that there's a conspiracy by scientists to fake photos of the earth and suppress their ideas.

- Apocalypticism: Another source of bizarre ideas comes from obsession with the end of the world and signs of the apocalypse. Ideas about the flips of the earth's magnetic field causing a global catastrophe fall in this category.

- Con artists and quacks: Some of the weird ideas we've discussed are clearly crooked con artists making a buck, getting publicity for their crank ideas, or just trying to swindle and fool people. The quacks who claim to predict earthquakes fall into this category, as do "healers" who promote "alternative medicine" with crystals and other pseudoscientific ideas and the dowsers with their fraudulent methods of finding water or other objects. The people behind the expanding earth also seem to be promoting their crank ideas to get attention and possibly make money off people who buy into their crazy notions. However, some of these people may not be conscious frauds and con artists; they may truly believe in the fakery that they are selling. This doesn't make their beliefs any less false.

- Urban myths: Many of the myths associated with earthquakes, especially earthquake weather and false ideas about how the earth moves, are simply urban myths that have been repeated often enough, especially in the uncritical media, to become widely known and accepted by many people.

The Echo Chamber

Another factor that explains the surprising prevalence of these weird ideas is the media and the internet. At one time, there were just a handful of TV and radio networks, and only three networks had national news shows on a regular basis. Thanks to the Fairness Doctrine, the major national networks (CBS, NBC, ABC) had to employ extensive fact-checking and had to try their best to avoid being too partisan in their coverage of events; they also had to label the more opinionate material as *editorial* in order to clearly distinguish between the factual news and the journalist's opinions.

Because they reported news to people of all political beliefs and were required by the Federal Communications Commission to follow the Fairness Doctrine of balancing their coverage of controversial topics, major media and their reporters strove for objectivity and made an effort to be truly "balanced and fair" when dealing with controversial issues—even to the point of giving someone who was clearly wrong at least some voice in the coverage. Newspapers, magazines, and TV news all employed fact-checkers who would track down the details of a story and make sure there was some external corroboration before they reported it. For a story involving some expertise to evaluate, reporters would routinely talk to numerous expert sources. Science reporting was often superficial and sometimes oversimplified, but at least the reporters talked to real scientists and strove to get a diversity of scientific expertise represented when they reported a scientific controversy.

Contrast that with the media landscape today. Thanks to the deregulation of the airwaves that the Reagan administration pushed through in the 1980s, the old Fairness Doctrine came to an end. No longer did TV news have to present both sides of a controversy in politics. A TV channel could be completely partisan with no effort to be fair and balanced. The deregulation of the airwaves also led to a huge number of new channels on TV, radio, and elsewhere. The age of the old mainstream media giants in print and broadcasting is rapidly dying, and subscriptions for print news and viewership for the major network news shows shrink every year. In their place are hundreds of different TV channels, including Fox News, which shamelessly calls itself "fair and balanced" even though it is the mouthpiece for the right-wingers in the United States and openly accepts money from right-wing causes and political organizations.

Fox News was founded by Rupert Murdoch explicitly to push his conservative probusiness agenda, as he does in the many other media he owns. Fox News was originally run by the late Roger Ailes, former Republican media adviser to Nixon, Reagan, and George H. W. Bush, who has repeatedly stated his belief that their coverage should reflect the right-wing viewpoint. As Ailes himself said, "'The truth' is whatever people will believe."[8] We also have MSNBC, which is famous for its left-wing commentators. Only CNN was relatively centrist in their coverage, but for a while they did a hard-right turn and tried to be a version of Fox and now are losing ground to Fox and MSNBC.

The explosion of additional media outlets in the 1980s and 1990s created a giant need for additional material to put on the air. Shows about Bigfoot, aliens and UFOs, ghosts, and paranormal beliefs like flat-eartherism, geocentrism, and creationism found a place in the legitimate media, where previously they would have been rejected from the airwaves as nonsense. UFO and alien shows are now a staple of the cable TV networks, especially formerly scientific channels. As I wrote elsewhere,

> It happens with disgusting regularity. You will flip through the various basic cable channels which are nominally "science oriented" (often grouped together on the dial if they feature scientific topics) and come up with nothing but junk, pseudoscience, and worse. "Reality shows" about subjects with little or no actual "reality" or science content, tons of paranormal and pseudoscientific shows promoting ghosts, UFOs, Bigfoot, and creationism—all fill the airwaves for channels like Discovery, The Learning Channel, History Channel, and even the Science Channel and National Geographic Channel. We watch a few minutes of these with complaints to anyone within earshot, then (usually) move on—or occasionally we get sucked in to watch the whole thing, like gawkers at a car crash. . . .
>
> So we all complain about the changes in our basic cable channels, and wonder why such dreck can make it on the air, but seldom think hard

about the process. But the excellent website TVTropes does a very nice job analyzing what happens to TV networks over time. To no one's surprise, it comes down to one simple factor: ratings (and therefore money from advertisers), largely driven by the effort to woo those big-spending trend-setting 18–31 male viewers who already dictate the movie industry's bottom line (although movies aim even lower to reach teenage boys, the biggest-spending and most loyal movie audience). As TVTropes points out (and those of us old enough to remember can attest to), it wasn't always this bad on cable TV. When the laws changed and the opportunity to create hundreds of basic cable channels first emerged in the 1980s, the channels were initially set up to fill specific programming niches, from the Golf Channel to the Game Show Network and so on. In the early 1980s, all these new niche-driven cable channels were very distinct and more or less true to their niche description. But since these are commercial channels that must sell ads based on numbers of viewers, the same factors that affect every other commercial enterprise came into play: keep tweaking it and give the customer whatever sells the most. (This dynamic does not apply to non-commercial stations like PBS in the U.S., or the BBC in Britain, which can program what they feel is in the public interest).

As TVTropes documents, nearly all these niche-defined networks have undergone "network decay" since they were founded in the 1980s, as their programming shifts to find hit shows. Because they are nearly all chasing nearly the same demographic of 18–31 year old males, they end up programming a lot of the same kinds of things (or even the same shows). Their original mission and distinctive programming are lost in a sea of reality shows and junk that keeps you in your seat, whether it be explosions or dangerous occupations or whatever. Another factor has been the expansion of media conglomerates, so that these multiple cable channels are owned by just a few corporations, and the CEO of each channel must answer to corporate bosses who are only interested in their profitability, not any abstract "mission" to air certain types of programming. So much for the high-minded idealism that drove the deregulation of the airwaves in the 1970s and 1980s, with the intent of offering us dozens of distinct choices. Instead, they all "decay" to a lowest-common-denominator of "if it bleeds, it ledes" bottom-line mentality, negating whatever real advantages that dozens of distinctive niche cable channels once offered. As TVTropes points out, the decisions are made by network execs worried only about their ratings and bottom lines, not any high-minded ideal like "quality television" that PBS brags so loudly about. They could (and did) notice that professional "wrestling" is popular with their 18–31 male demographic, and see no problem with programming the WWE next to a show about science.[9]

Finally, the biggest factor that promotes the spread of many of these weird ideas is the giant cesspool of lies we call the internet. Before the internet became widely available, there were always a small number of cranks and crazies who believed in the flat earth or geocentrism, but their ability to reach other like-minded people was very limited. They tried to publish small journals or newsletters and maintain mailing lists, but communication among all these people dispersed widely around the world was difficult. Even more importantly, they were not easy to find, so someone who might be curious about the flat earth or moon-landing conspiracy ideas had to try very hard to locate other people promoting the ideas. There were a few shows like Art Bell's late-night talk show on *Coast to Coast* radio, which brought the crazies and conspiracy nuts out of the woodwork, exposing their ideas to other like-minded people (and to critics), but the reach of this radio community was limited.

Now, thanks to the internet, these people can type a few search terms into their browser and instantly find websites for each of these crank ideas. They can quickly read a whole host of websites full of bad information and see convincing but misleading or fake YouTube videos. They can join virtual communities of like-minded individuals and soon have all their bad ideas confirmed. Unlike scientific literature, which has peer review to weed out bad information, or old mainstream media of newspapers and news stations with their fact-checkers, anyone can put any garbage on the internet.

Many young people who have grown up with the internet and surf the web a lot have learned to spot bad information as they become more cautious. Still, a lot of garbage gets too much attention, because people are gullible and people like to hear their own beliefs and prejudices confirmed. Sadly, I do not think this situation is going to get any better as long as the internet remains a cesspool of "fake news" and false information and there is no fact-checking, peer review, or quality control. Instead, as the polls discussed below show, more and more people are trusting fake sources on the internet and beginning to reject expert opinion about complicated topics in favor of their own gut instincts about what is true.

Science, Our Candle in the Darkness

As discussed in the first chapter, science has a built-in mechanism of rooting out bad ideas: peer review. Scientists are human, they aren't perfect, and they can be misled by their own biases and ideologies, but in most cases, the harsh scrutiny of other scientists soon weeds out the bad data and gives us at least some standard for deciding if an idea has merit. Scientists are not immune to cultural forces of the world around them, but by and large, they are not openly ideological either. Most of the ones I know are largely nonpolitical and are appalled when they see other scientists bias their work to suit political or ideological ends. Most of them still follow the concept of a scientific reality that must be respected, no matter what their biases. There are, of course, a few examples of scientific bias and fraud, but pointing them out simply highlights how rare and unusual they really are, the exceptions to the rule. In most instances, the scientific community does a good job of policing itself and trying to winnow out what is real from what is not.

More to the point, scientists often discover things that go against our belief systems, but they must put aside their favorite ideas and face this reality. When Copernicus and Galileo showed us that the earth (and humankind) is not in the center of the universe, it wasn't popular, but it was true. Everyone except a handful of geocentrists and the uneducated now look at the sun "rising" and "setting" and have come to terms with the counterintuitive notion that it is the earth turning instead. When Darwin showed that life had evolved and that we are all closely related to other living things, not specially created, it offended many people (and still does), but its truth was soon acknowledged by the entire scientific community and nearly all educated Westerners who weren't religiously biased, even before Darwin died.

As we discussed in chapter 1, science is our "candle in the darkness," to use Carl Sagan's term. Our modern world, with all its benefits, is almost entirely product of scientific advances. Most people respect and listen to science when it comes to practical issues like engineering or medicine that improve their lives. In fact, in our gut, we respect science so much that many people try to imitate it and put on its trappings and appearance, even pseudoscientists like the creationists or quack doctors who are pushing garbage in the name of science. Science is practically the only form of thought or scholarship that is self-checking and relatively immune to intellectual fads, because ultimately scientific ideas have to prove their merit and meet their own form of reality check when they are tested against the real world. The fact that science doesn't always tell us what we want to hear is further proof that we cannot impose our biases or ideologies on it and still be practicing scientists.

In this context, most people regard flat-earthers, geocentrists, and Young-Earth Creationists as harmless cranks, more amusing than threatening. But there are signs that these rejections of science are serious threats to society as a whole. If the general ignorance of Americans is not shocking enough, their ignorance of science is even more staggering. Study after study over the years have shown a virtually unchanging and abysmally poor understanding of how the world really works.[10] These include such howlers as:

- Only 53 percent of adults know how long it takes for the earth to revolve around the sun.

- Only 59 percent of adults know that dinosaurs and humans never coexisted (the "Flintstones model of prehistory").

- Only 47 percent of adults can guess correctly the percentage of the earth's surface covered by water.

- Only 21 percent of adults answered all three of the previous questions correctly.

- And a surprisingly large number of American adults still think the sun revolves around the earth!

This is not just the crackpot fanatics from the Galileo Was Wrong geocentrist website,[11] people who believe this out of pure biblical literalism. No one knows how many American adults even think the earth is flat, but it's probably a lot more than just the crazies who have officially joined the flat-earth movement.

A shockingly large number of adults do not know which is larger, an electron or an atom. Most adults cannot give simple definitions of concepts like the cell, the molecule, or DNA. Only about 33 percent of adults agree with the notion that more than half of human genes are identical to those of mice, and only 38 percent of adults realize that humans share 98 percent of their DNA with chimpanzees.[12] Only 35 percent think the Big Bang describes the early history of our universe. Carl Sagan estimated that 95 percent of American adults were scientifically illiterate.[13] Sagan was thinking of a far higher level of science literacy than these simple middle-school-level science-knowledge questions we have just mentioned, and judging from the numbers we have just cited, he was not far off.

If American adults are so appallingly illiterate in science, what about teenagers who are still supposed to be taking science classes in school? Sadly, the numbers are just as depressing. Most kids of high school age know about the same amount of science or less than adults who haven't sat in a high school science class for years. According to a study by Jon Miller of Northwestern University, an expert on science literacy, US high school students are "below average and below most European countries" on virtually every academic achievement test administered in the past thirty years. Miller found that exposure to a college science course made significant improvements on science literacy but only as measured against a baseline of almost total ignorance.[14] Currently, scholars are studying the concept of civic science literacy, which is more than just knowledge of science facts; it means understanding science well enough to apply it to everyday lives. Here again, the results are equally depressing. Although the numbers are slowly rising, in a 2007 study, Miller found that the civic science literacy of Americans was still less than 30 percent. As Miller put it, "We should take no pride in a finding that 70 percent of Americans cannot read and understand the science section of the *New York Times*."[15]

Another way to frame the question is to ask how we stack up against other countries. Study after study has shown that the United States is near the bottom of the major industrialized nations in science literacy. One recent study found that among fifteen-year-olds, the United States ranked twenty-ninth among the nations of the world.[16] At the top of the list was Finland, followed by a number of other northern European countries (the other Scandinavian countries, Germany, France, the United Kingdom, plus developed or developing Asian countries like Japan, South Korea, and China). Nearly every other ranking in recent years gives similar results. Although the exact order of the top ten countries might be shuffled a bit, the United States always comes out near the bottom, along with countries like Turkey and Cyprus that have a fraction of our wealth and our spending on education. That alone is a mark of disgrace for our society—that we can spend so much money per child and yet end up with such miserable results that nearly every other industrialized country does far better than we do. What does that say for our future economic well-being when we're near the bottom on crucial things like understanding science?

Some recent studies show that the picture is not getting better but worse, especially with millennials and the younger generation just finishing their educations.[17] One study showed that only 66 percent of eighteen- to twenty-four-year-olds are confident that the world is round; 4 percent actually were flat-earthers, and the rest had some doubts about it. Nine percent said that they once believed the earth was round but were beginning to have doubts, while 5 percent said they thought the world was flat but were having doubts. The rest (16 percent) just weren't sure. For comparison, 94 percent of those fifty-five and older think the world is round, as do 85 percent of forty-five- to fifty-four-year-olds, 82 percent of thirty-five- to forty-four-year-olds, and 76 percent of twenty-five- to thirty-four-year-olds. A 2019 study by Chapman University found that 57% of Americans believe in Atlantis and 41% believe in alien visitation, up from only 21% and 40% in 2016.

Another survey by YouGov found that 84 percent of Americans believe the world is round, 5 percent believe it is round but are becoming skeptical, 2 percent said the world is flat, 2 percent said the world was flat but were having doubts, and 7 percent weren't sure.[18] Not surprisingly, the main predictor of flat-earth beliefs was religion, since a majority of those people who reported themselves as flat-earthers were very religious. As we saw in chapter 2, biblical literalism is the main reason for the beliefs of most of the active flat-earth organizations.

Another way to gauge the acceptance of weird ideas is to see how their presence on the internet is trending through time. As this article on Live Science points out, "Google Trends suggests that interest in the concept of 'flat Earth,' if not necessarily belief, has been on the rise over the past few years. The search trend for the term in the United States crept upward over 2016 and 2017, with spikes coinciding with particular events. For example, searches for 'flat Earth' rose around the time of the August 2017 solar eclipse, which spurred much sparring between flat-Earthers and mainstreamers online."[19]

Even if the hard-core version of geocentrism that we saw in chapter 3 isn't a popular notion, widespread scientific illiteracy means that lots of people who are not hard-core geocentrists are confused by their muddled memories of their educations; they just don't understand the difference between saying that the earth revolves around the sun and vice versa. A 1999 Gallup poll found that 18 percent of Americans mistakenly said the sun revolves around the earth, and this was true of similar percentages of Germans and Britons as well.[20]

It wasn't always this bad. During the *Sputnik* scare and space race of the late 1950s and early 1960s, Americans were shocked to discover how far they had fallen behind and brought back rigorous and engaging science education, only to see it languish as pseudoscience (especially creationism) has eaten away at the textbooks. The demands of standardized testing have pushed the curriculum toward subjects covered on the test, leaving science (and physical education, art, music, and many other subjects) in the cold.

Thanks to the scientific illiteracy of our general population and the hostile environment for science that pseudoscience is providing, we are falling behind many other nations in the one area we used to excel in: science and technology. Study after study has documented a brain drain of scientists going to other countries with less anti-intellectualism and more favorable climates for science, especially in fields like stem-cell research or cloning that offend the right-wingers who currently run the country. America can no longer compete to make the cheapest or best electronics, toys, cars, or most anything else, as China, Korea, India, Singapore, Indonesia, and many other nations have taken those tasks away from us in corporate cutbacks and outsourcing. For many years, we could brag that we won a lion's share of the Nobel Prizes in science, but that dominance is now coming to an end as well. If we can't compete with other nations in manufacturing and commerce and we give away our advantages in science and technology, what kind of world are we leaving for our children? What does this imply for our national security, when we farm out not only blue-collar but also white-collar jobs and then are slaves to other countries that are doing better science and technology as well?

As Carl Sagan put it in his 1996 book *The Demon-Haunted World*: "We've arranged a global civilization in which the most critical elements profoundly depend on science and technology. We have also arranged things so that almost no one understands science and technology. This is a prescription for disaster."[21] As Stephen Hawking put it, "In a democracy, it is very important that the public have a basic understanding of science so that they can control the way that science and technology increasingly affect our lives."[22]

A report by British education professor Harry T. Dyer provided a window on how pseudoscientists think. He attended the 2018 Flat Earth Conference in Birmingham, England, and made some interesting observations. In his words,

> There was also a lot of team-building, networking, debating, workshops—and scientific experiments.
>
> Yes, flat earthers do seem to place a lot of emphasis and priority on scientific methods and, in particular, on observable facts. The weekend in no small part revolved around discussing and debating science, with lots of time spent running, planning, and reporting on the latest set of flat earth experiments and models. Indeed, as one presenter noted early on, flat earthers try to "look for multiple, verifiable evidence" and advised attendees to "always do your own research and accept you might be wrong."
>
> While flat earthers seem to trust and support scientific methods, what they don't trust is scientists, and the established relationships between "power" and "knowledge." This relationship between power and knowledge has long been theorised by sociologists. By exploring this relationship, we can begin to understand why there is a swelling resurgence of flat earthers. . . . What's important here is not necessarily whether they believe the earth is flat or not, but instead what their resurgence and public conventions tell us about science and knowledge in the 21st century.
>
> Multiple competing models were suggested throughout the weekend, including "classic" flat earth, domes, ice walls, diamonds, puddles with multiple worlds inside, and even the earth as the inside of a giant cosmic egg. The level of discussion however often did not revolve around the models on offer, but on broader issues of attitudes towards existing structures of knowledge, and the institutions that supported and presented these models.[23]

Dyer goes on to point out that what we're seeing is an erosion of trust, not only in political institutions and the media but in science as well. Years of attacks by right-wing politicians against the science of evolution and climate change have done a lot of damage to the reputation of science and scientists. Lots of people no longer look up to science to provide answers but believe that their own gut instincts, or whatever internet "research" they find, is sufficient to decide the truth for themselves. As Dyer argues, "The age of the expert may be passing. Now, everybody has the power to create and share content. When Michael Gove, a leading proponent of Brexit, proclaimed: 'I think the people of this country have had enough of experts,' it would seem that he, in many ways, meant it."[24]

Although Dyer's examples are mostly from Britain (for example, most experts predicted Brexit would be an economic disaster, yet a slim majority of Britons voted for it), his point applies to the United States as well. We see a Trump administration and a Republican Party that defy expertise in science across the board. They are not only rejecting the overwhelming worldwide consensus on climate change but are also aggressively pursuing policies to destroy the environment so that their big business backers will be able to pollute and develop and mine without restrictions.

As Dyer described the debates at the 2018 Flat Earth Convention,

> Despite early claims, from as far back as HG Wells' "world brain" essays in 1936, that a worldwide shared resource of knowledge such as the internet would create peace, harmony and a common interpretation of reality, it appears that quite the opposite has happened. With the increased voice afforded by social media, knowledge has been increasingly decentralized, and competing narratives have emerged.
>
> This was something of a reoccurring theme throughout the weekend, and was especially apparent when four flat earthers debated three physics PhD students. A particular point of contention occurred when one of the physicists pleaded with the audience to avoid trusting YouTube and bloggers. The audience and the panel of flat earthers took exception to this, noting that "now we've got the internet and mass communication . . . we're not reliant on what the mainstream are telling us in newspapers, we can decide for ourselves." It was readily apparent that the flat earthers were keen to separate knowledge from scientific institutions.
>
> At the same time as scientific claims to knowledge and power are being undermined, some power structures are decoupling themselves from scientific knowledge, moving towards a kind of populist politics that are increasingly skeptical of knowledge. This has, in recent years, manifested itself in extreme ways—through such things as public politicians showing support for Pizzagate or Trump's suggestions that Ted Cruz's father shot JFK.

But this can also be seen in more subtle and insidious form in the way in which Brexit, for example, was campaigned for in terms of gut feelings and emotions rather than expert statistics and predictions. Science is increasingly facing problems with its ability to communicate ideas publicly, a problem that politicians, and flat earthers, are able to circumvent with moves towards populism.

Again, this theme occurred throughout the weekend. Flat earthers were encouraged to trust "poetry, freedom, passion, vividness, creativity, and yearning" over the more clinical regurgitation of established theories and facts. Attendees were told that "hope changes everything," and warned against blindly trusting what they were told. This is a narrative echoed by some of the celebrities who have used their power to back flat earth beliefs, such as the musician B.O.B, who tweeted: "Don't believe what I say, research what I say."

In many ways, a public meeting of flat earthers is a product and sign of our time; a reflection of our increasing distrust in scientific institutions, and the moves by power-holding institutions towards populism and emotions. In much the same way that Foucault reflected on what social outcasts could reveal about our social systems, there is a lot flat earthers can reveal to us about the current changing relationship between power and knowledge. And judging by the success of this UK event—and the large conventions planned in Canada and America this year—it seems the flat earth is going to be around for a while yet.[25]

That is why in most of the chapters I have tried to not only describe the weird ideas of flat-earthers, geocentrists, hollow earthers, moon-landing conspiracists, and Young-Earth Creationists but also to outline the scientific evidence for *why* scientists accept certain ideas about the earth and universe and reject others. It's clear that telling students that the earth is round or the sun is the center of the solar system, as most kids in K–12 grades learn by rote, isn't enough. Ideally, science education should also describe our *evidence why* we accept these ideas as true. That's asking a lot for K–12 education, where everything is focused on rote learning and a lot of multiple choice tests. But certainly at the college level, where I teach, it should be standard practice, especially in science classes that are for nonmajors to learn some real science.

The late great Carl Sagan said it best:

There's another reason I think popularizing science is important, why I try to do it. It's a foreboding I have—maybe ill-placed—of an America in my children's generation, or my grandchildren's generation, when all the manufacturing industries have slipped away to other countries; when we're a service and information-processing economy; when awesome techno-logical powers are in the hands of a very few, and no one representing the public interest even grasps the issues; when the people (by "the people" I mean the broad population in a democracy) have lost the ability to set their own agendas, or even to knowledgeably question those who do set the agendas; when there is no practice in questioning those in authority; when, clutching our crystals and religiously consulting our horoscopes, our critical faculties in steep decline, unable to distinguish between what's true and what feels good, we slide, almost without noticing, into supersti-tion and darkness.[26]

NOTES

Chapter 1

1. Bill Nye, *Undeniable: Evolution and the Science of Creation* (New York: St. Martin's Press, 2014), 183.

2. Prachi Gupta, "Neil deGrasse Tyson: Science Is True 'Whether or Not You Believe in It,'" *Salon*, March 11, 2014, https://www.salon.com/2014/03/11/neil_degrasse_tyson _science_is_true_whether_or_not_you_believe_in_it/.

3. Ann Finkbeiner, "Abstruse Goose: Piss 'Em Off," Last Word on Nothing, https:// www.lastwordonnothing.com/2014/01/21/abstruse-goose-piss-em-off/?fbclid =IwAR1a_PdecC0K79KSOAVGbs5RySMxdPClhpS7ZABQIuRhxFLYupShxL3XsoA.

4. George Santayana, *The Life of Reason* (New York: Charles Scribner's Sons, 1905), 10.

5. Bertrand Russell, *History of Western Philosophy* (New York: Simon and Schuster, 2008), 527.

6. Carl Sagan, *Broca's Brain: Reflections on the Romance of Science* (New York: Random House, 1979), 12.

7. Carl Sagan, *Cosmos*, episode 4, "Heaven and Hell," KCET, aired October 19, 1980.

8. Richard Feynman, *Surely You're Joking, Mr. Feynman!* (New York: Norton, 1997), 343.

9. Richard P. Feynman, *Feynman Lectures on Physics*, vol. 3 (Reading, MA: Addison Wesley, 1989), 8–12.

10. Thomas Henry Huxley, *Aphorisms and Reflections from the Works of Thomas Henry Huxley* (London: Macmillan, 1908), 4.

11. Sunil Laxman, "Crosstalk: Understanding Counterintuitive Science Needs a Culture of Rigorous Scepticism," *Wire*, May 20, 2016, https://thewire.in/history/crosstalk -understanding-counterintuitive-science-needs-a-culture-of-rigorous-sceptisicm.

12. "Tertullian on the Flesh of Christ," The Tertullian Project, http://www.tertullian.org/.

13. *The Spiritual Exercises of St. Ignatius of Loyola*, trans. Father Elder Mullan (New York: P. J. Kenedy & Sons, 1914).

14. Sagan, *Broca's Brain*, 117.

15. Arthur Schopenhauer, *The World as Will and Representation* (Leipzig: Brodhaus, 1818).

16. John Gray, *Seven Types of Atheism* (New York: Farrar Straus Giroux, 2015), 111.

17. William Saletan, "Inside the Minds of the JFK Conspiracy Theorists," *New Scientist*, November 22, 2013, https://www.newscientist.com/article/dn24626-inside-the-minds -of-the-jfk-conspiracy-theorists/.

18. Tom Jensen, "Democrats and Republicans Differ on Conspiracy Theory Beliefs," Public Policy Polling, April 2, 2013, http://www.publicpolicypolling.com/pdf/2011/PPP _Release_National_ConspiracyTheories_040213.pdf.

19. "Zogby America Likely Voters 8/23/07 thru 8/27/07 MOE +/- 3.1 percentage points," 911truth.org, accessed October 7, 2019, http://www.911truth.org/images/Zogby Po112007.pdf.

20. Jensen, "Democrats and Republicans Differ."

Chapter 2

1. Huffington Post Media, "New 'View' Co-Host Sherri Shepherd Doesn't Know If World Is Flat," *Huffington Post*, March 28, 2008, https://www.huffingtonpost. com/2007/09/18/new-view-cohost-sherri-sh_n_64864.html.

2. Harrison Wind, "'Just Walk Outside and Use Your Five Senses': Wilson Chandler Backs Up Irving's Flat Earth Theory," DNVR, February 18, 2017, http://www.bsndenver .com/just-walk-outside-and-use-your-five-senses-wilson-chandler-backs-up-irvings-flat -earth-theory/.

3. Sopan Deb, "Kyrie Irving Doesn't Know If the Earth Is Round or Flat. He Does Want to Discuss It," *New York Times*, June 8, 2018, https://www.nytimes.com/2018/06/08 /movies/kyrie-irving-nba-celtics-earth.html.

4. "Shaquille O'Neal Comes Out for Flat Earth—Interviewer Stunned—Shaq," You-Tube video, 2:45, posted by "markksargent," March 18, 2017, https://www.youtube.com /watch?v=bGVtC52XjIg.

5. Kyle Tasman, "Stefon Diggs Takes to Twitter to Share Flat Earth Beliefs," 247 Sports, February 18, 2017, https://247sports.com/nfl/minnesota-vikings/ContentGallery/Stefon -Diggs-flat-earth-twitter-Minnesota-Vikings—51359057/.

6. Emma Baccellieri, "Kyrie Irving Really, Actually, Earnestly Believes the Earth Is Flat," Deadspin, February 18, 2017, https://deadspin.com/kyrie-irving-really-actually -earnestly-believes-the-e-1792511889.

7. Emmett Knowlton, "Shaquille O'Neal Recants His Stance on the Earth Being Flat: 'I'm Joking, You Idiots,'" Business Insider, March 24, 2017, https://www.businessinsider .com/shaquille-o-neal-flat-earth-conspiracy-im-joking-you-idiots-2017-3.

8. Kristian Winfield, "Kyrie Irving Apologizes to Science Teachers for Spreading Flat Earth Theories," SBNation, October 1, 2018, https://www.sbnation.com/2018/10/1 /17925768/kyrie-irving-apology-flat-earth-theory-celtics.

9. Laura Wagner, "Neil DeGrasse Tyson Gets into a Rap Battle with B.o.B over Flat Earth Theory," NPR, January 26, 2016, https://www.npr.org/sections/thetwo-way/2016 /01/26/464474518/neil-degrasse-tyson-gets-into-a-rap-battle-with-b-o-b-over-flat-earth -theory.

10. Wagner, "Neil DeGrasse Tyson Gets into a Rap Battle with B.o.B."

11. Wagner, "Neil DeGrasse Tyson Gets into a Rap Battle with B.o.B."

12. Trevor Nace, "Flat Earth Rocket Man Finally Blasts Off in Homemade Rocket to Prove Earth Is Flat," *Forbes*, March 27, 2018, https://www.forbes.com/sites/trevornace /2018/03/27/flat-earth-rocket-man-finally-blasts-off-in-homemade-rocket-to-prove-earth -is-flat/#4077179f9b6f.

13. Pat Graham, "Self-Taught Rocket Scientist Plans to Launch over Ghost Town," AP News, November 21, 2017, https://apnews.com/9d8e5e8e9245412ab80f5a1f58d885b7.

14. Wagner, "Neil DeGrasse Tyson Gets into a Rap Battle with B.o.B."

15. Phil Plait, "Flat Earth? Really?" *Discover*, August 11, 2018, http://blogs.discover magazine.com/badastronomy/2008/08/11/flat-earth-really/#.XBm2Qs9KhBw.

16. *Plato in Twelve Volumes,* vol. 9, trans. W. R. M. Lamb (Cambridge: Harvard University Press, 1925), 53.

17. John Sacrobosco, *Treatise on a Sphere*, chapter I.9, http://www.esotericarchives.com /solomon/sphere.htm.

18. Christine Garwood, *Flat Earth: History of an Infamous Idea* (New York: Macmillan, 2007).

19. Bible Study Tools Staff, "Bible Verses about Flat Earth," Bible Study Tools, May 22, 2018, https://www.biblestudytools.com/topical-verses/bible-verses-about-flat-earth/; Robert J. Schadewald, "The Flat-Earth Bible," accessed October 7, 2019, https://www .lockhaven.edu/~dsimanek/febible.htm.

20. Robert J. Schadewald, "Six 'Flood' Arguments Creationists Can't Answer," NCSE, accessed October 7, 2019, https://ncse.com/cej/3/3/six-flood-arguments-creationists-cant -answer.

21. Eddy Gilmore, "The Earth Is Not Only Flat—It's Motionless, Too," *Cincinnati Enquirer*, March 26, 1967, https://www.newspapers.com/clip/17484882/the_cincinnati _enquirer/.

22. Douglas Martin, "Charles Johnson, 76, Proponent of Flat Earth," *New York Times*, March 25, 2001, https://www.nytimes.com/2001/03/25/us/charles-johnson-76-proponent -of-flat-earth.html; Robert J. Schadewald. "The Flat-Out Truth: Earth Orbits? Moon Landings? A Fraud! Says This Prophet," Flat Earth Society, https://www.theflatearthsociety.org /library/newspaperandmagazine/Flat-Out%20Truth,%20The%20(Schadewald).pdf.

23. The Flat Earth Society, accessed October 7, 2019, https://www.theflatearthsociety .org/home/.

24. Stephanie Pappas, "What in the World? Flat-Earthers Gather at First Conference," Live Science, November 17, 2017, https://www.livescience.com/60972-flat-earthers-first -conference.html.

25. Harry T. Dyer, "I Watched an Entire Flat Earth Convention—Here's What I Learned," Live Science, May 8, 2018, https://www.livescience.com/62506-flat-earth -convention.html.

26. Stephanie Pappas, "A Third of Young Millennials Are Confused about This Incontrovertible Fact," Live Science, April 4, 2018, https://www.livescience.com/62220-millennials-flat-earth-belief.html?fbclid=IwAR0TqKNV84HGpE06ARsbrCBj9qgPiPntHMCTU fz6Afx9XXSwkczH7hlQOjo.

27. Paul Ratner, "Reality Show Idea: Make Flat-Earthers Search for the World's Edge," Big Think, September 25, 2018, https://bigthink.com/surprising-science/reality-show -about-flat-earthers?rebelltitem=3#rebelltitem3.

28. Andrew Whalen, "'Behind the Curve' Ending: Flat Earthers Disprove Themselves with Own Experiments in Netflix Documentary," Newsweek, February 25, 2019, https:// www.newsweek.com/behind-curve-netflix-ending-light-experiment-mark-sargent -documentary-movie-1343362?fbclid=IwAR2n12N38rGlgGdcYeubJg5ZqxOgH5HYbso3j d2SC36d7BpmVLs64pj-dpU.

29. Whalen, "'Behind the Curve' Ending."

30. Whalen, "'Behind the Curve' Ending."

31. Whalen, "'Behind the Curve' Ending."

32. Rob Waugh, "Why Are Flat Earthers Wrong? This Astrophysicist Believes He Can Explain the Answer," Yahoo News, March 13, 2018, https://www.yahoo.com/news /astrophysicist-neil-degrasse-tyson-explains-flat-earthers-wrong-140858722.html.

33. Ethan Siegel and Starts with a Bang, "Five Impossible Facts That Would Have to Be True If the Earth Were Flat," Forbes, November 24, 2017, https://www.forbes.com/sites /startswithabang/2017/11/24/five-impossible-facts-that-would-have-to-be-true-if-the -earth-were-flat/?fbclid=IwAR3oAw8bataTDSny-DLx91uRgEvoK_ENfElyeqjcqkc9eI T95VoArI9dHks#160a1a297c4f.

34. Aristotle, De Caelo, 298a, 2–10.

35. "Pan Am Flight 50," YouTube video, 21:45, posted by "FanofPanAm," March 2, 2013, https://www.youtube.com/watch?v=5Ci4G2JGxQo.

36. "Polar Route," Wikimedia, accessed October 7, 2019, https://upload.wikimedia.org /wikipedia/commons/1/1d/PolarRoute.png.

37. "Do Any Flights Go over the South Pole?" Aviation, accessed October 7, 2019, https://aviation.stackexchange.com/questions/33938/do-any-flights-go-over-the-south -pole.

38. "Do Any Flights Go Over the South Pole?"

39. University of Leicester, "Students Film Breathtaking Curvature of Earth Using High-Altitude Weather Balloon," Phys.org, January 6, 2017, https://phys.org/news/2017 –01-students-breathtaking-curvature-earth-high-altitude.html.

Chapter 3

1. Robert Steinback, "Anti-Semitic 'Radical Traditionalist' Catholics Insist the Universe Revolves around the Earth. The Devil Reportedly Disagrees," SPLC, February 23, 2011, https://www.splcenter.org/fighting-hate/intelligence-report/2011/geocentrism -%E2%80%98seminar%E2%80%99-hosted-radical-traditionalist-catholics.

2. Galileo Was Wrong, accessed October 7, 2019, http://galileowaswrong.com/.

3. Ginger Thompson, "Immigration on Tap for Bush, Fox Talks," *New York Times*, March 26, 2006, https://news.google.com/newspapers?id=_1kaAAAAIBAJ&sjid =XCYEAAAAIBAJ&dq=robert-sungenis&pg=6714,4991566.

4. Colin Lecher, "The Conspiracy Theorist Who Duped the World's Biggest Physicists," *Popular Science*, May 7, 2014, https://www.popsci.com/article/science/how -conspiracy-theorist-duped-worlds-biggest-physicists.

5. Box Office Mojo, accessed October 7, 2019, https://www.boxofficemojo.com /movies/?id=principle.htm.

6. Lecher, "Conspiracy Theorist."https://www.popsci.com/article/science/how -conspiracy-theorist-duped-worlds-biggest-physicists

7. Sigmund Freud, "Lecture XVIII: Fixation to Traumas—the Unconscious," in *Lectures Introducing Psychoanalysis* (London: George Allen & Unwin, 1917).

8. Steinback, "Anti-Semitic 'Radical Traditionalist.'"

9. "Rotating Earth from Space (Galileo Spacecraft 1990)," YouTube video, 2:21, posted by "Films for the Earth," February 3, 2010, https://www.youtube.com/watch?v =UVuqcEuIRgs.

Chapter 4

1. Dante Alighieri, *Inferno*, Canto III, lines 1–9, Mandelbaum translation: http://www .worldofdante.org/comedy/dante/inferno.xml/1.3.

2. Will Storr, "Hollow Earth Conspiracy Theories: The Hole Truth," *Telegraph*, July 13, 2014, https://www.telegraph.co.uk/culture/books/10961412/Hollow-Earth-conspiracy -theories-the-hole-truth.html.

3. Storr, "Hollow Earth Conspiracy Theories."

4. Gregory L. Reece, *UFO Religion: Inside Flying Saucer Cults and Culture* (London: I. B. Tauris, 2007), 17.

5. Storr, "Hollow Earth Conspiracy Theories."

6. Storr, "Hollow Earth Conspiracy Theories."

7. Storr, "Hollow Earth Conspiracy Theories."

8. Staff Writer, "Actor Dustin Hoffman Lobbies for More Reality in Science-Fiction Movies," News.com, August 4, 2009, https://www.news.com.au/news/actor-dustin -hoffman-lobbies-for-more-reality-in-science-fiction-movies/news-story/d0ce7b96 1a7da975ec01b9b5ca74d892.

Chapter 5

1. "Neal Adams—Science: 01—Conspiracy: Earth Is Growing!" YouTube video, 10:01, posted by "nealadamsdotcom," March 2, 2007, http://www.youtube.com /watch?v=0JfBSc6e7QQ.

2. "Expanding Earth My Ass," YouTube video, 9:02, posted by "potholer54," November 3, 2010, http://www.youtube.com/watch?v=epwg60d49e8.

3. M. W. McElhinny, S. R. Taylor, and D. J. Stevenson, "Limits to the Expansion of Earth, Moon, Mars, and Mercury and to Changes in the Gravitational Constant," *Nature* 271, no. 5643 (1978): 316–21; P. W. Schmidt and D. A. Clark, "The Response of Palaeomagnetic Data to Earth Expansion," *Geophysical Journal of the Royal Astronomical Society* 61 (1980): 95–100.

4. NASA/Jet Propulsion Laboratory, "It's a Small World, after All: Earth Is Not Expanding, NASA Research Confirms," Science Daily, August 17, 2011, http://www.sciencedaily.com/releases/2011/08/110817120527.htm.

5. X. Wu, X. Collilieux, Z. Altamimi, B. L. A. Vermeersen, R. S. Gross, and I. Fukumori, "Accuracy of the International Terrestrial Reference Frame Origin and Earth Expansion," *Geophysical Research Letters* 38, no. L13304 (2011): 1–5.

6. G. E. Williams, "Geological Constraints on the Precambrian History of Earth's Rotation and the Moon's Orbit," *Reviews of Geophysics* 38, no. 1 (2000): 37–59.

7. A. E. Beck, "Energy Requirements for an Expanding Earth," *Journal of Geophysical Research* 66, no. 5 (1961): 1485–90.

8. S. Van Der Lee and G. Nolet, "Seismic Image of the Subducted Trailing Fragments of the Farallon Plate," *Nature* 386, no. 6622 (1997): 266.

9. "Expanding Earth My Ass."

10. Beck, "Energy Requirements."

11. "Earth Expansion: Main Scientific Evidence," dinox.org, accessed October 7, 2019, https://www.dinox.org/expandingearth.html.

12. "Earth Expansion: Main Scientific Evidence."

Chapter 6

1. Andrew Chaikin, *A Man on the Moon* (New York: Penguin, 1994).

2. Rogier van Bakel, "The Wrong Stuff," *Wired*, January 9, 1994, https://www.wired.com/1994/09/moon-land/?pg=5.

3. Robert Scheaffer, *Psychic Vibrations: Skeptical Giggles from the Skeptical Inquirer* (Charleston, SC: James Randi Educational Foundation, 2011), 12–15.

4. Gallup News Service, "Did Men Really Land on the Moon?" Gallup, February 15, 2001, https://news.gallup.com/poll/1993/Did-Men-Really-Land-Moon.aspx.

5. Philip C. Plait, *Bad Astronomy: Misconceptions and Misuses Revealed, from Astrology to the Moon Landing "Hoax"* (New York: Wiley, 2002), 155–73.

6. Seth Borenstein, "Book to Confirm Moon Landings," Knight Ridder Newspapers, November 2, 2002, https://web.archive.org/web/20090726105516/http://archive.deseretnews.com/archive/946348/Book-to-confirm-moon-landings.html.

7. Phil Plait, "Moon Hoax +10," *Discover*, February 15, 2011, http://blogs.discovermagazine.com/badastronomy/2011/02/15/moon-hoax-10/#.XCgdVs9KhBw.

8. http://bd.fom.ru/report/cat/sci_sci/kosmos/of001605.

9. Piers Bizony, "It Was a Fake, Right?," E&T, July 6, 2009, https://web.archive.org/web/20110128204607/http://eandt.theiet.org/magazine/2009/12/fake-right.cfm.

10. "Moon Landing Hoax," *Mythbusters*, accessed October 7, 2019, https://www .dailymotion.com/video/x2m7k1z.

11. Clavius.org, accessed October 7, 2019, http://www.clavius.org/envrad.html.

12. Mike Bara, "Who Mourns for Apollo? Part II," Studyphysics.com, accessed October 7, 2019, http://www.studyphysics.ca/apol102.pdf.

13. "Apollo 15 Hammer and Feather Drop," YouTube video, 1:22, posted by "Stop and Think," April 27, 2009, https://www.youtube.com/watch?v=-4_rceVPVSY.

14. Plait, *Bad Astronomy*, 163–64.

15. "Apollo 16—Sample 61016 Collection," YouTube video, posted by "Toby Smith," February 18, 2016, https://www.youtube.com/watch?v=9jeEtFEMP2I.

16. Jim Longuski, *The Seven Secrets of How to Think Like a Rocket Scientist* (New York: Springer, 2006).

Chapter 7

1. Terrence Aym, "Magnetic Pole Shift May Destroy Civilization," Before It's News, November 20, 2012, https://beforeitsnews.com/v3/science-and-technology/2012/2496818 .html.

2. "Magnetic Pole Shifting Happening Now," Global Rumblings, November 18, 2012, https://globalrumblings.blogspot.com/2012/11/magnetic-pole-shifting-happening-now .html.

3. "Magnetic Reversals," Poleshift.com, accessed October 7, 2019, http://poleshift.com /magnetic-reversals.

4. David Montaigne, "Chaos and Cover-Ups: What Evidence Exists of an Ancient Pole Shift?" Ancient Origins, April 19, 2018, https://www.ancient-origins.net/news-science -space/chaos-and-cover-ups-evidence-exists-ancient-pole-shift-009921.

5. J. D. Hays, "Faunal Extinctions and Reversal of the Earth's Magnetic Field," *Geological Society of America Bulletin* 82 (1971): 2433–47.

6. Peter Sonnenfeld, "Effects of a Variable Sun at the Beginning of the Cenozoic Era," *Climate Change* 1, no. 4 (1978): 355–382, https://link.springer.com/article/10.1007 %2FBF00135156#page-1; Roy E. Plotnick, "Relationship between Biological Extinctions and Geomagnetic Reversals," *Geology* 8, no. 12 (1980): 578–81, https://pubs .geoscienceworld.org/gsa/geology/article-abstract/8/12/578/187677/relationship-between -biological-extinctions-and?redirectedFrom=fulltext.

Chapter 8

1. Charles Darwin, *Voyage of the Beagle* (New York: Penguin, 1989), chapter 16.

2. K. A. Borden and S. L. Cutter, "Spatial Patterns of Natural Hazards Mortality in the United States," *International Journal of Health Geographics* 7 (2008): 64.

3. W. J. Humphreys, *Monthly Weather Review*, 180–181. War Department, Office of the Chief Signal Officer, 1919.

4. Snopes Staff, "Earthquake Myths," Snopes, May 10, 2010, https://www.snopes.com /fact-check/earthquake-myths/#eCGrQsgCLj9Tp5HL.99.

5. University of Miami Rosenstiel School of Marine and Atmospheric Science, "Link between Earthquakes and Tropical Cyclones: New Study May Help Scientists Identify Regions at High Risk for Earthquakes," Science News, December 26, 2011, https://www .sciencedaily.com/releases/2011/12/111208121016.htm; Richard A. Lovett, "Hurricane May Have Triggered Earthquake Aftershocks," Nature, April 19, 2013, https://www.nature .com/news/hurricane-may-have-triggered-earthquake-aftershocks-1.12839.

Chapter 9

1. "California Quake Predicted—March 19th," eBaum's World, March 18, 2011, http:// www.ebaumsworld.com/videos/california-quake-predicted-march-19th/81380209/.

2. US Geological Survey, "The Loma Prieta Earthquake of October 17, 1989," accessed October 7, 2019, https://pubs.usgs.gov/unnumbered/70039527/report.pdf.

3. R. Hunter, "Can Jim Berkland Predict Earthquakes?" Skeptical Inquirer 30, no. 5 (2006): 47–50.

4. M. Kennedy, J. E. Vidale, and M. G. Parker, "Earthquakes and the Moon: Syzygy Predictions Fail the Test," Seismological Research Letters 75, no. 5 (2004): 607–12.

5. S. Hartzell and T. Heaton, "The Fortnightly Tide and the Tidal Triggering of Earthquakes," Bulletin of the Seismological Society of America, 79 (1989): 1282–86; J. E. Vidale, D. C. Agnew, M. J. S. Johnston, and D. H. Oppenheimer, "Absence of Earthquake Correlation with Earth Tides: An Indication of High Preseismic Fault Stress Rate," Journal of Geophysical Research 103 (1998): 24567–72.

6. M. Tolstoy, F. A. Vernon, J. A. Orcutt, and F. K. Wyatt, "Breathing of the Seafloor: Tidal Correlations of Seismicity at Axial Volcano," Geology 30 (2002): 503–6.

7. Brian Dunning, "Animal Earthquake Prediction," Skeptoid, October 23, 2018, https://skeptoid.com/episodes/4646; Heiko Woith, Gesa M. Petersen, Sebastian Hainzl, and Torsten Dahm, "Review: Can Animals Predict Earthquakes?" Bulletin of the Seismo- logical Society of America 108, no. 3A (2018): 1031–45, https://pubs.geoscienceworld .org/ssa/bssa/article-abstract/108/3A/1031/530275/review-can-animals-predict -earthquakes-review-can?redirectedFrom=fulltext.

8. Lee Dye, "Ability to Forecast Quakes Shaky at Best, Experts Say," LA Times, Novem- ber 18, 1990, https://www.latimes.com/archives/la-xpm-1990-11-18-mn-6880-story .html.

9. Aidan Lewis, "Row over Italian Quake 'Forecast,'" BBC News, April 6, 2009, http:// news.bbc.co.uk/2/hi/europe/7986585.stm; "L'Aquila Quake: Italy Scientists Guilty of Man- slaughter," BBC News, October 22, 2012, https://www.bbc.com/news/world-europe -20025626.

10. Lewis, "Row over Italian Quake 'Forecast.'"

11. Kenneth Chang, "Earthquakes' Many Mysteries Stymie Efforts to Predict Them," New York Times, April 13, 2009, https://www.nytimes.com/2009/04/14/science /14quak.html?mtrref=www.skepticblog.org&gwh=538C6EFABACF11098D009544429 60178&gwt=pay.

12. Nick Squires quoted in L'Aquila and Gordon Rayner, "Italian Earthquake: Expert's Warnings Were Dismissed as Scaremongering," *Telegraph*, April 6, 2009, https://www.telegraph.co.uk/news/worldnews/europe/italy/5114139/Italian-earthquake-experts-warnings-were-dismissed-as-scaremongering.html.

13. Thomas H. Jordan, "Don't Blame Italian Seismologists for Quake Deaths," *New Scientist*, September 21, 2011, https://www.newscientist.com/article/mg21128310-200-dont-blame-italian-seismologists-for-quake-deaths/.

14. "L'Aquila Quake: Scientists See Convictions Overturned," *BBC News*, November 10, 2014, https://www.bbc.com/news/world-europe-29996872.

15. "L'Aquila Quake."

16. "L'Aquila Quake."

Chapter 10

1. St. Augustine, *The Literal Interpretation of Genesis 1:19–20*, chap. 19.

2. Ronald Numbers, *The Creationists* (New York: Knopf, 1992), 3.

3. Numbers, *Creationists*, 39.

4. Numbers, *Creationists*, 39.

5. Numbers, *Creationists*, 40.

6. Numbers, *Creationists*, 89–101.

7. Numbers, *Creationists*, 125.

8. Numbers, *Creationists*, 125.

9. Numbers, *Creationists*, 190.

10. Henry Morris, *Biblical Cosmology and Modern Science* (Nutley, NJ: Craig Press, 1970), 32–33.

11. Stephen Jay Gould, "Genesis vs. Geology," *Atlantic*, September 1982, https://www.theatlantic.com/magazine/archive/1982/09/genesis-vs-geology/306198/.

12. "Conspiracy Road Trip: Creationism," YouTube video, 57:12, posted by "Tr3Ve10cita," October 8, 2012, https://www.youtube.com/watch?v=Oju_lpqa6Ug.

Chapter 11

1. Christians Against Dinosaurs, accessed October 7, 2019, https://www.christiansagainstdinosaurs.com/.

2. CADbot, "Christian Parent Thrown off Mumsnet for Anti-Dinosaur Rant," *Huffington Post*, posted on Christians Against Dinosaurs, October 18, 2016, https://www.christiansagainstdinosaurs.com/christian-parent-thrown-off-mumsnet-for-anti-dinosaur-rant-huffington-post/; Oliver McAteer, "Christian Tries to Wipe Out Dinosaurs from Classrooms in Bizarre Mumsnet Post after Child 'Bites Three Kids on Face,'" *Metro News*, February 12, 2015, https://metro.co.uk/2015/02/12/christian-tries-to-wipe-out-dinosaurs-from-classrooms-in-bizarre-mumsnet-post-after-child-bites-three-kids-on-face-5059551/?ito=v-a.

3. "Dinosaur Hoax: Ask a Christian," YouTube video, 7:18, posted by "Christians Against Dinosaurs," December 7, 2014, https://www.youtube.com/watch?v=eXlLos6f7Go.

4. Ryan Barrell, "Christian Parent Thrown off Mumsnet for Anti-Dinosaur Rant," *Huffington Post*, February 12, 2015, https://www.huffingtonpost.co.uk/2015/02/12/dinosaurs-mumsnet-burn-dinos_n_6670934.html?ec_carp=8526721666937132021.

5. McAteer, "Christian Tries to Wipe Out Dinosaurs."

6. "Poe's Law," Rationalwiki.org, accessed October 7, 2019, https://rationalwiki.org/wiki/Poe's_Law.

7. "Poe's Law," Wikipedia, accessed December 15, 2019, https://en.wikipedia.org/wiki/Poe%27s_law.

8. Galileo Was Wrong, accessed October 7, 2019, http://galileowaswrong.com/.

9. "Ra-Men Special—Dinosaur Deniers—with Kristen Auclair," YouTube video, 1:05:10, posted by "AronRa," February 18, 2015, https://www.youtube.com/watch?v=72M-f4BhCGQ.

10. "Full Text of 'The Musket,'" Archive.org, accessed October 7, 2019, https://archive.org/stream/musket2004sout/musket2004sout_djvu.txt.

11. "Christians Against Dinosaurs Is a Hoax," Uncertaintist, February 24, 2015, https://uncertaintist.wordpress.com/2015/02/24/christians-against-dinosaurs-is-a-hoax/.

12. "Christians Against Dinosaurs—CAD Facebook page," accessed October 7, 2019, https://www.facebook.com/ChristiansAgainstDinosaurs.

13. "Dinosaurs: Science or Science Fiction," accessed October 7, 2019, http://www.ocii.com/~dpwozney/dinosaurs.htm.

14. "Dinosaurs."

15. "Are Dinosaurs Real?" YouTube video, 9:52, posted by "TheWordprophet," February 7, 2013, https://www.youtube.com/watch?v=rllQpnWrVBQ.

16. Morgan Matthew, "Bill Nye Boo'd in Texas for Saying the Moon Reflects the Sun," Think Atheist, February 21, 2009, http://www.thinkatheist.com/profiles/blogs/bill-nye-bood-in-texas-for.

Chapter 12

1. Paul Braterman, "Kelvin, Rutherford, and the Age of the Earth: I, the Myth," 3 Quarks Daily, January 27, 2014, https://www.3quarksdaily.com/3quarksdaily/2014/01/kelvin-rutherford-and-the-age-of-the-earth-i-the-myth.html.

2. Cherry Lewis, *The Dating Game: One Man's Search for the Age of the Earth* (Cambridge: Cambridge University Press, 2012), 57.

3. Lewis, *Dating Game*, 58.

4. Cited in Lewis, *Dating Game*, 111.

Chapter 13

1. John Muir, "Letters, 1874–1888, of a Personal Nature, about Mount Shasta," in *The Life and Letters of John Muir*, vol. II, ed. William Frederic Bade (New York: Houghton Mifflin, 1923), 29.

2. College of the Siskiyous Library, accessed December 15, 2019, https://www.siskiyous.edu/library/shasta/mscollection.htm.

3. Theodore Roosevelt, "Letter to Harrie Cassie Best, dated Nov. 12, 1908, White House," in *Harry Cassie Best: Painter of the Yosemite Valley, California Oaks, and California Mountains*, ed. George Wharton James (1930), 18.

4. Ted Neild, *Supercontinent: Ten Billion Years in the Life of Our Planet* (Cambridge, MA: Harvard University Press, 2007), 38–9.

5. "The Origin of the Lemurian Legend," Archive.org, accessed October 7, 2019, https://web.archive.org/web/20120919063057/http://www.siskiyous.edu/shasta/fol/lem/index.htm.

6. Mark Moran and Mark Sceurman, *Weird California* (San Francisco: Sterling, 2006), 58.

7. "Harmonic Convergence," Archive.org, accessed October 7, 2019, https://web.archive.org/web/20100527184532/http://www.siskiyous.edu/shasta/fol/har/index.htm.

8. Shareen Strauss, "UFO Conference Message: 'Light, Love, Life,'" Mtshastanews.com, August 1, 2018, http://www.mtshastanews.com/news/20180801/ufo-conference-message-light-love-life.

9. "UFOs Filmed at Mt Shasta," YouTube video, 10:16, posted by "ET Connections," August 12, 2018, https://www.youtube.com/watch?v=BBKXzfXbuZg.

10. "Breaking News: UFO Hangar Door Opens on Top of Mt. Adams/ECETI June 30 2017," YouTube video, 9:35, posted by "FADE TO BLACK Radio," July 9, 2017, https://www.youtube.com/watch?v=fepbvdws9Xk.

11. "Mt. Shasta—The False Door," YouTube video, 9:17, posted by "Right Hemispheric Remote Viewing," March 2, 2017, https://www.youtube.com/watch?v=1-qizyYXPmE.

12. "Is There a Secret Underground City in Mount Shasta?" YouTube video, 7:20, posted by "Paranormal Junkie," March 30, 2018, https://www.youtube.com/watch?v=M4tbbGgrCCo.

13. http://www.siskiyous.edu/shasta/fol/har/index.htm.

14. http://www.alienskyships.com/UFO_clouds.html.

15. "Mount Shasta," US Geological Survey, accessed October 7, 2019, https://volcanoes.usgs.gov/volcanoes/mount_shasta/mount_shasta_monitoring_4.html.

Chapter 14

1. Plato, "Timaeus," accessed October 7, 2019, http://classics.mit.edu/Plato/timaeus.html.

2. "Critias by Plato," Project Gutenberg, accessed October 7, 2019, http://www.gutenberg.org/ebooks/1571.

3. Rodney Castleden, *Atlantis Destroyed* (London: Routledge, 2001).

4. T. Taylor, trans. and ed., *Proclus: Proclus' Commentary on the Timaeus of Plato*, Thomas Taylor series 15–16 (Frome: Prometheus Trust, 1998; first edition 1816), 117.10–30.

5. Abraham Ortelius, "Gadiricus," in *Thesaurus Geographicus* (Antwerp: Plantin, 1596).

6. J. Annas, *Plato: A Very Short Introduction* (Oxford: Oxford University Press, 2003), 42.

7. Kenneth L. Feder, *Frauds, Myths, and Mysteries: Science and Pseudoscience in Archaeology* (Mountain View, CA: Mayfield, 1999), 86.

8. Quoted in Feder, *Frauds, Myths, and Mysteries*, 88.

9. A. Giovannini, "Peut-on démythifier l'Atlantide?" *Museum Helveticum* 42 (1985): 151–56.

10. S. Soter and D. Katsonopoulou, "Submergence and Uplift of Settlements in the Area of Helike, Greece, from the Early Bronze Age to Late Antiquity," *Geoarchaeology* 26, no. 4 (2011): 584.

Chapter 15

1. Richard Muir, *Woods, Hedgerows and Leafy Lanes* (Chalford: Tempus, 2008), 163.

2. Aimé Michel, *Mystérieux objects célestes* [About flying saucers] (Paris: Arthaud, 1958).

3. idoubtit, "Leylines: From the Old Straight Track to the Ghostbusters Vortex," Spooky Geology, February 20, 2017, https://spookygeology.com/leylines-from-the-old -straight-track-to-the-ghostbusters-vortex/.

4. idoubtit, "Leylines."

5. "Ley Lines and the Connection to Adverse Spiritual Phenomena," *Supernatural Magazine*, accessed October 7, 2019, http://supernaturalmagazine.com/articles/ley-lines -and-the-connection-to-adverse-spiritual-phenomena.

6. Alfred Watkins, *The Old Straight Track: Its Mounds, Beacons, Moats, Sites, and Mark Stones* (London: Methuen & Co., 1925).

7. Watkins, *Old Straight Track*.

8. Alfred Watkins, "Early British Trackways, Moats, Mounds, Camps and Sites," Internet Sacred Text Archive, 1922, http://www.sacred-texts.com/neu/eng/ebt/index.htm.

9. G. H. Piper, "Arthur's Stone, Dorstone," *Transactions of the Woolhope Naturalists' Field Club* 1881–82 (1888): 175–80.

10. Cited in Nigel Pennick and Paul Devereux, *Lines on the Landscape: Leys and Other Linear Enigmas* (London: Hale, 1989), 88.

11. Paul Devereux and Ian Thomson, *The Ley Hunter's Companion: Aligned Ancient Sites: A New Study with Field Guide and Maps* (London: Thames and Hudson, 1979).

12. John Michell, *City of Revelation: On the Proportions and Symbolic Numbers of the Cosmic Temple* (London: Sphere, 1973).

13. Gerald of Wales and John William Sutton, "The Tomb of King Arthur," Camelot Project, 2001, https://d.lib.rochester.edu/camelot/text/gerald-of-wales-arthurs-tomb.

14. Michael Shermer, *The Believing Brain: From Ghosts and Gods to Politics and Conspiracies: How We Construct Beliefs and Reinforce Them as Truths* (New York: Times Books, 2011).

15. Matthew Johnson, *Archaeological Theory: An Introduction*, 2nd ed. (Malden, MA: Wiley-Blackwell, 2009), 5.

16. David G. Kendall, "A Survey of the Statistical Theory of Shape," *Statistical Science* 4, no. 2 (May 1989): 83–89.

17. Clive L. N. Ruggles, "Ley Lines," in *Ancient Astronomy: An Encyclopaedia* (Santa Barbara: ABC-CLIO, 2005), 225.

18. "Pizzalines8.png," Wikimedia Commons, October 7, 2019, https://commons .wikimedia.org/wiki/File:Pizzalines8.png.

19. Matt Parker, "Aliens with a Taste for Pick 'n' Mix: Woolworths Stores Follow Uncanny Geometrical Patterns," *Times Online*, August 1, 2010, https://web.archive.org/web /20100120112826/http://timesonline.typepad.com/science/2010/01/aliens-with-a-taste -for-pick-n-mix-woolworths-stores-follow-uncanny-geometrical-patterns.html.

Chapter 16

1. Mindvalley Academy, "The Secrets of Crystal Healing: A Complete Guide to Supercharging the Mind, Body and Spirit with Sacred Stones and Minerals," *Conscious Lifestyle Mag*, accessed October 7, 2019, https://www.consciouslifestylemag.com/crystal-healing -guide/.

2. Mindvalley Academy, "Secrets of Crystal Healing."

3. Rinku Patel, "The Only 7 Crystals You Need to Boost Your Mood + Live Your Best Life," mbgmindfulness, accessed October 7, 2019, https://www.mindbodygreen.com/0 –20957/the-only-7-crystals-you-need-to-boost-your-mood-live-your-best-life.html.

4. Patel, "Only 7 Crystals You Need to Boost Your Mood."

5. "Crystal Healing," Answers.com, accessed October 7, 2019, http://www.answers .com/search?q=crystal+healing.

6. "The Science of Crystal Healing," India Reiki Master, accessed October 7, 2019, http://www.indianreikimasters.com/crystalheal.html.

7. Elizabeth Palermo, "Crystal Healing: Stone-Cold Facts about Gemstone Treatments," Live Science, June 23, 2017, https://www.livescience.com/40347-crystal-healing .html.

8. "Crystal Woo," Rationalwiki.org, accessed October 7, 2019, https://rationalwiki.org /wiki/Crystal_woo.

Chapter 17

1. Samuel Sheppard, *Epigrams Theological, Philosophical, and Romantick* (London: Bucknell, 1651).

2. Georgius Agricola, *De Re Metallica*, trans. Herbert Hoover (New York: Dover, 1950), 38.

3. "Witching for Water," Spooky Geology, accessed December 14, 2019, https:// spookygeology.com/witching-for-water/.

4. Agricola, *De Re Metallica*, 82.

5. William Pryce, *Mineralogica Cornubiensis* (New York: Gale ECCO, 1778).

6. W. F. Barrett, "Physical Research," Archive.org, accessed October 7, 2019, https://archive.org/stream/psychicalresearoobarr#page/n5/mode/2up.

7. "Witching for Water," Spooky Geology.

8. "Witching for Water," Spooky Geology.

9. J. W. Gregory, "Water Divining," Archive.org, accessed October 7, 2019, https://archive.org/stream/annualreportofbo1928smit#page/n381/mode/2up.

10. W. A. MacFadyen, "Some Water Divining in Algeria," *Nature* 157 (1946): 304–405.

11. M. J. Aitken, "Test for Correlation between Dowsing Response and Magnetic Disturbance," *Archaeometry* 2 (1959): 58–9.

12. R. A. Foulkes, "Dowsing Experiments," *Nature* 229 (1971): 163–8.

13. J. G. Taylor and E. Balanovski, "Can Electromagnetism Account for Extra-Sensory Phenomena?" *Nature* 276, no. 5683 (1978): 64–7.

14. Evon Z. Vogt and Ray Hyman, *Water Witching U.S.A.*, 2nd ed. (Chicago: University of Chicago Press, 1979).

15. Martijn Van Leusen, "Dowsing and Archaeology," *Archaeological Prospection* 5 (1998): 123–38.

16. William E. Whittaker, "Grave Dowsing Reconsidered," Office of the State Archaeologist, Iowa, accessed October 7, 2019, https://archaeology.uiowa.edu/sites/archaeology.uiowa.edu/files/Dowsing.pdf.

17. Robert Konig, Jurgen Moll, and Armadeo Sarma, "The Kassel Dowsing Test: Part 1," Geotech, accessed October 7, 2019, http://www.geotech1.com/cgi-bin/pages/common/index.pl?page=lrl&file=info/kassel/kasse11.dat.

18. H. Wagner, H.-D. Betz, and H. L. König, "Schlußbericht 01 KB8602," Bundesministerium für Forschung und Technologie, 1990.

19. J. T. Enright, "Testing Dowsing: The Failure of the Munich Experiments," *Skeptical Inquirer* 23, no. 1 (1999), https://www.csicop.org/si/show/testing_dowsing_the_failure_of_the_munich_experiments.

20. Sandia National Laboratories, "Double-Blind Field Evaluation of the MOLE Programmable Detection System," https://web.archive.org/web/20091104173209/http://www.justnet.org/Lists/JUSTNET%20Resources/Attachments/440/moleeval_apr02.pdf.

21. Rod Nordland, "Iraq Swears by Bomb Detector U.S. Sees as Useless," *New York Times*, November 3, 2009, https://www.nytimes.com/2009/11/04/world/middleeast/04sensors.html?_r=1&ref=world&mtrref=en.wikipedia.org&gwh=6BAFE3D8966FC13E3B99534DFB1341FF&gwt=pay.

22. "Test Report: The Detection Capability of the Sniffex Handheld Explosives Detector," Naval Sea Systems Command, accessed October 7, 2019, http://s3.amazonaws.com/propublica/assets/docs/NavyReport.pdf.

23. Dan Rivers, "Tests Show Bomb Scanner Ineffective, Thailand Says," CNN, February 16, 2010, http://www.cnn.com/2010/WORLD/asiapcf/02/16/thailand.bomb.scanner/index.html?eref=rss_topstories.

24. "2.5.3.8 EXPRAY Field Test Kit," National Criminal Justice Reference Service, accessed October 7, 2019, https://www.ncjrs.gov/pdffiles1/nij/178913–2.pdf.

Chapter 18

1. Marquis de Condorcet, *Sketch for a Historical Picture of the Progress of the Human Spirit* (Paris: Agasse, 1795).

2. Condorcet, *Sketch*.

3. Michael Shermer, *The Believing Brain: From Ghosts and Gods to Politics and Conspiracies: How We Construct Beliefs and Reinforce Them as Truths* (New York: Times Books, 2011).

4. Matt Taibbi, *The Great Derangement: A Terrifying True Story of War, Politics, and Religion* (New York: Spiegel & Grau, 2008).

5. Shermer, *Believing Brain*, 22.

6. "The Backfire Effect: Why Facts Don't Win Arguments," Big Think, October 15, 2013, https://bigthink.com/think-tank/the-backfire-effect-why-facts-dont-win-arguments.

7. Bruno Maddox, "Blinded by Science," *Discover Magazine*, February 2007, 28–29.

8. "Fox News Controversies," Wikipedia, October 7, 2019, http://en.wikipedia.org/wiki/Fox_News_Channel_controversies.

9. Donald R. Prothero, "Science TV 'Network Decay,'" Skeptic Blog, January 25, 2012, http://www.skepticblog.org/2012/01/25/science-tv-sell-out/.

10. News Staff, "Science Literacy—American Adults 'Flunk' Basic Science, Says Survey," Science20.com, March 11, 2009, http://www.science20.com/news_releases/science_literacy_american_adults_flunk_basic_science_says_survey; Liza Gross, "Scientific Illiteracy and the Partisan Takeover of Biology," Curious Cat Blog, June 17, 2006, http://engineering.curiouscatblog.net/2006/06/17/scientific-illiteracy/.

11. Galileo Was Wrong, accessed October 7, 2019, http://galileowaswrong.blogspot.com/.

12. http://news.msu.edu/story/1087/.

13. Carl Sagan, *The Demon-Haunted World: Science as a Candle in the Dark* (New York: Ballantine, 1996).

14. Rachel Ehrenberg, "Latest Science Survey Is Heavy on Trivia, Light on Concepts," Science News, September 17, 2015, https://www.sciencenews.org/blog/culture-beaker/latest-science-survey-heavy-trivia-light-concepts.

15. Michigan State University, "Scientific Literacy: How Do Americans Stack Up?" Science Daily, February 27, 2007, http://www.sciencedaily.com/releases/2007/02/070218134322.htm.

16. Amanda Paulson, "New Report Ranks U.S. Teens 29th in Science Worldwide," *Christian Science Monitor*, December 5, 2007, http://www.csmonitor.com/2007/1205/p02s01-usgn.html.

17. Stephanie Pappas, "A Third of Young Millennials Are Confused about This Incontrovertible Fact," Live Science, April 4, 2018, https://www.livescience.com/62220

-millennials-flat-earth-belief.html?fbclid=IwAR0TqKNV84HGpE06ARsbrCBj9qgPi PntHMCTUfz6Afx9XXSwkczH7hlQOjo.

18. Hoang Nguyen, "Most Flat Earthers Consider Themselves Very Religious," YouGov, April 2, 2018, https://today.yougov.com/topics/philosophy/articles-reports/2018/04/02 /most-flat-earthers-consider-themselves-religious.

19. Pappas, "Third of Young Millennials Are Confused."

20. Steve Crabtree, "New Poll Gauges Americans' General Knowledge Levels," Gallup, July 6, 1999, https://news.gallup.com/poll/3742/new-poll-gauges-americans-general -knowledge-levels.aspx.

21. Sagan, *Demon-Haunted World*, 6.

22. Stephen W. Hawking, 2011, "There Is No Heaven; It's a Fairy Story," *The Guardian*, May 15, 2011.

23. Harry T. Dyer, "I Watched an Entire Flat Earth Convention—Here's What I Learned," Live Science, May 8, 2018, https://www.livescience.com/62506-flat-earth -convention.html.

24. Dyer, "Flat Earth Convention."

25. Dyer, "Flat Earth Convention."

26. Sagan, *Demon-Haunted World*, 403.

FOR FURTHER READING

Chapter 1

Dunning, Brian. 2018. *Conspiracies Declassified: The Skeptoid Guide to the Truth Behind the Theories*. New York: Adams Media.

Gardner, Martin. 1952. *Fads and Fallacies in the Name of Science*. New York: Dover.

———. 1981. *Science: Good, Bad, and Bogus*. Buffalo, NY: Prometheus.

Goldacre, Ben. 2010. *Bad Science: Quacks, Hacks, and Big Pharma Flacks*. New York: McClelland and Stewart.

Gorham, Geoffrey. 2012. *Philosophy of Science: A Beginner's Guide*. New York: OneWorld.

Loxton, Daniel, and Donald R. Prothero. 2013. *Abominable Science!: Origins of the Yeti, Nessie, and Other Famous Cryptids*. New York: Columbia University Press.

Popper, Karl. 1935. *The Logic of Scientific Discovery*. London: Routledge Classics.

———. 1963. *Conjectures and Refutations: The Growth of Scientific Knowledge*. London: Routledge Classics.

Prothero, Donald R., and Timothy D. Callahan. 2018. *UFOs, Chemtrails and Aliens: What Science Says*. Bloomington: Indiana University Press.

Ruggiero, Vincent Ryan. 2011. *Beyond Feelings: A Guide to Critical Thinking*. New York: McGraw-Hill.

Sagan, Carl. 1996. *The Demon-Haunted World: Science as a Candle in the Dark*. New York: Ballantine.

Schick, Theodore Jr., and Lewis Vaughn. 2013. *How to Think about Weird Things: Critical Thinking for a New Age*. New York: McGraw-Hill.

Shermer, Michael. 1997. *Why People Believe Weird Things: Pseudoscience, Superstition, and Other Confusions of Our Time*. New York: W. H. Freeman.

———. 2005. *Science Friction: Where the Known Meets the Unknown*. New York: Times Books.

———. 2011. *The Believing Brain: From Ghosts and Gods to Politics and Conspiracies—How We Construct Beliefs and Reinforce Them as Truth*. New York: Times Books.

West, Mick. 2018. *Escaping the Rabbit Hole: How to Debunk Conspiracy Theories Using Facts, Logic, and Respect*. New York: Skyhorse.

Chapter 2

Gardner, Martin. 1952. *Fads and Fallacies in the Name of Science*. New York: Dover.

———. 1981. *Science: Good, Bad, and Bogus*. Buffalo, NY: Prometheus.

Garwood, Christine. 2008. *Flat Earth: The History of an Infamous Idea*. New York: Thomas Dunne.

Russell, Jeffrey Burton. 1991. *Inventing the Flat Earth: Columbus and Modern Historians*. New York: Praeger.

West, Mick. 2018. *Escaping the Rabbit Hole: How to Debunk Conspiracy Theories Using Facts, Logic, and Respect*. New York: Skyhorse.

Chapter 3

Copernicus, Nicholas. (1543) 1995. *On the Revolutions of Heavenly Spheres*. Buffalo, NY: Prometheus.

Galilei, Galileo. (1632) 2001. *Dialogue Concerning the Two Chief World Systems: Ptolemaic and Copernican*. New York: Modern Library.

Gardner, Martin. 1952. *Fads and Fallacies in the Name of Science*. New York: Dover.

———. 1981. *Science: Good, Bad, and Bogus*. Buffalo, NY: Prometheus.

Heilbron, John. 2010. *Galileo*. Oxford: Oxford University Press.

Keating, Karl. 2015. *The New Geocentrists*. New York: Rasselas House.

Shea, William R., and Mariano Artigas. 2003. *Galileo in Rome: The Rise and Fall of a Troublesome Genius*. Oxford: Oxford University Press.

Wootton, David. 2010. *Galileo: Watcher of the Skies*. New Haven: Yale University Press.

Chapter 4

Brown, Geoffrey. 1993. *The Inaccessible Earth: An Integrated View of Its Structure and Composition*. Berlin: Springer.

Gardner, Martin. 1952. *Fads and Fallacies in the Name of Science*. New York: Dover.

———. 1981. *Science: Good, Bad, and Bogus*. Buffalo, NY: Prometheus.

Poirier, Jean-Paul. 2000. *Introduction to the Physics of the Earth's Interior*. Cambridge: Cambridge University Press.

Prothero, Donald R. 2018. *The Story of the Earth in 25 Rocks*. New York: Columbia University Press.

Reece, Gregory L. 2007. *UFO Religion: Inside Flying Saucer Cults and Culture*. London: I. B. Tauris.

Terasaki, Hidenori, and Rebecca A. Fischer. 2016. *Deep Earth: Physics and Chemistry of the Lower Mantle and Core*. Washington, DC: American Geophysical Union.

Chapter 5

Beck, A. E. 1961. "Energy Requirements for an Expanding Earth." *Journal of Geophysical Research* 66, no. 5: 1485–90.

Carey, S. Warren. 1975. *The Expanding Earth*. Amsterdam: Elsevier.

McElhinny, M. W., S. R. Taylor, and D. J. Stevenson. 1978. "Limits to the Expansion of Earth, Moon, Mars, and Mercury and to Changes in the Gravitational Constant." *Nature* 271, no. 5643: 316–21.

Prothero, Donald R. 2018. *The Story of the Earth in 25 Rocks.* New York: Columbia University Press.

Schmidt, P. W., and D. A. Clark. 1980. "The Response of Palaeomagnetic Data to Earth Expansion." *Geophysical Journal of the Royal Astronomical Society* 61: 95–100.

Van Der Lee, S., and G. Nolet. 1997. "Seismic Image of the Subducted Trailing Fragments of the Farallon Plate." *Nature* 386, no. 6622: 266.

Williams, G. E. 2000. "Geological Constraints on the Precambrian History of Earth's Rotation and the Moon's Orbit." *Reviews of Geophysics* 38, no. 1: 37–59.

Wu, X., X. Collilieux, Z. Altamimi, B. L. A. Vermeersen, R. S. Gross, and I. Fukumori. 2011. "Accuracy of the International Terrestrial Reference Frame Origin and Earth Expansion." *Geophysical Research Letters* 38, L13304: 1–5.

Chapter 6

Chaikin, Andrew. 1994. *A Man on the Moon.* New York: Penguin.

Charles River Editors. 2013. *Apollo 11: The History and Legacy of the First Moon Landing.* Self-published, CreateSpace.

Donovan, James. 2019. *Shoot for the Moon: The Space Race and the Extraordinary Voyage of Apollo 11.* New York: Little Brown and Company.

Gibson, Philip. 2014. *#Houston 69: Apollo 11—When Men Walked on the Moon.* Self-published, CreateSpace.

Launius, Roger. 2019. *Apollo's Legacy: Perspectives on the Moon Landings.* Washington, DC: Smithsonian.

Plait, Philip C. 2002. *Bad Astronomy: Misconceptions and Misuses Revealed, from Astrology to the Moon Landing "Hoax."* New York: John Wiley & Sons.

Pyle, Rod, and Buzz Aldrin. 2019. *First on the Moon: The Apollo 11 50th Anniversary Experience.* New York: Sterling.

Scheaffer, Robert. 2011. *Psychic Vibrations: Skeptical Giggles from the Skeptical Inquirer.* Charleston, SC: James Randi Educational Foundation.

Whitehouse, David. 2019. *Apollo 11: The Inside Story.* New York: Icon.

Chapter 7

Butler, Robert F. 1991. *Paleomagnetism: Magnetic Domains to Geologic Terranes.* London: Blackwell.

Cox, Allan, ed. 1973. *Plate Tectonics and Geomagnetic Reversals.* San Francisco, CA: W. H. Freeman.

Hays, J. D. 1971. "Faunal Extinctions and Reversal of the Earth's Magnetic Field." *Geological Society of America Bulletin* 82: 2433–47.

McElhinny, Michael W., and Phillip L. McFadden. 1999. *Paleomagnetism: Continents and Oceans.* San Diego, CA: Academic Press.

Merrill, Ronald T. 2011. *Our Magnetic Earth: The Science of Geomagnetism.* Chicago: University of Chicago Press.

Mitchell, Alanna. 2018. *The Spinning Magnet: The Electromagnetic Force That Created the Modern World—and Could Destroy It*. New York: Dutton.

Prothero, Donald R. 2018. *The Story of the Earth in 25 Rocks*. New York: Columbia University Press.

Tauxe, Lisa. 2010. *Essentials of Paleomagnetism*. Berkeley: University of California Press.

Turner, Gillian. 2011. *North Pole, South Pole: The Epic Quest to Solve the Great Mystery of Earth's Magnetism*. New York: McGraw-Hill.

Chapters 8 and 9

Bolt, Bruce A. 2006. *Earthquakes*. New York: W. H. Freeman.

Collier, M. 1999. *A Land in Motion: California's San Andreas Fault*. Berkeley: University of California Press.

Dvorak, J. 2014. *Earthquake Storms: The Fascinating History and Volatile Future of the San Andreas Fault*. New York: Pegasus.

Fountain, Henry. 2017. *The Great Quake: How the Biggest Earthquake in North America Changed Our Understanding of This Planet*. New York: Crown.

Hough, Susan E. 2004. *Finding Fault in California: An Earthquake Tourist's Guide*. Missoula, MT: Mountain Press.

Jones, Lucy. 2018. *Big Ones: How Natural Disasters Have Shaped Us (and What We Can Do about Them)*. New York: Doubleday.

Miles, Kathryn. 2017. *Quakeland: On the Road to America's Next Devastating Earthquake*. New York: Dutton.

Winchester, Simon. 2006. *A Crack in the Edge of the World: America and the Great California Earthquake of 1906*. New York: Harper Perennial.

Yeats, Robert S., Kerry E. Sieh, and Clarence R. Allen. 1997. *Geology of Earthquakes*. Oxford: Oxford University Press.

Chapter 10

Beus, Stanley, and Michael Morales, eds. 1990. *Grand Canyon Geology*. Oxford: Oxford University Press.

Dalley, Stephanie. 2009. *Myths from Mesopotamia: Creation, the Flood, Gilgamesh, and Others*. Oxford: Oxford University Press.

Eldredge, Niles. 1982. *The Monkey Business: A Scientist Looks at Creationism*. New York: Pocket.

——. 2000. *The Triumph of Evolution and the Failure of Creationism*. New York: W. H. Freeman.

Friedman, Richard. 1987. *Who Wrote the Bible?* New York: Harper & Row.

Frye, Roland Mushat, ed. 1983. *Is God a Creationist?: The Religious Case Against Creation-Science*. New York: Charles Scribner.

Heidel, Alexander. 1942. *The Babylonian Genesis*. Chicago: University of Chicago Press.

——. 1946. *The Gilgamesh Epic and Old Testament Parallels*. Chicago: University of Chicago Press.

Isaak, Mark. 2006. *The Counter-Creationism Handbook*. Berkeley: University of California Press.

Kitcher, Philip. 1982. *Abusing Science: The Case Against Creationism*. Cambridge, MA: MIT Press.

McGowan, Chris. 1984. *In the Beginning: A Scientist Shows Why the Creationists Are Wrong*. Buffalo, NY: Prometheus.

Montgomery, David R. 2012. *The Rocks Don't Lie: A Geologist Investigates Noah's Flood*. New York: W. W. Norton.

Numbers, Ronald. 1992. *The Creationists: The Evolution of Scientific Creationism*. New York: Knopf.

Prothero, Donald R. 1990. *Interpreting the Stratigraphic Record*. New York: W. H. Freeman.

———. 2017. *Evolution: What the Fossils Say and Why It Matters*. 2nd ed. New York: Columbia University Press.

———. 2018. *The Story of the Earth in 25 Rocks*. New York: Columbia University Press.

Prothero, Donald R., and Fred Schwab. 2013. *Sedimentary Geology*. 3rd ed. New York: W. H. Freeman.

Sarna, Nahum. 1966. *Understanding Genesis: The Heritage of Biblical Israel*. New York: Schocken.

Chapter 11

Allison, Peter A., and David J. Bottjer, eds. 2011. *Taphonomy: Process and Bias through Time*. Berlin: Springer Verlag.

———. 1991. *Taphonomy: Releasing Data Locked in the Fossil Record*. Berlin: Springer Verlag.

Behrensmeyer, Anna K. 1980. *Fossils in the Making: Vertebrate Taphonomy and Paleoecology*. Chicago: University of Chicago Press.

Donovan, Steven. 1991. *The Process of Fossilization*. New York: Columbia University Press.

Montgomery, David R. 2012. *The Rocks Don't Lie: A Geologist Investigates Noah's Flood*. New York: W. W. Norton.

Prothero, Donald R. 1990. *Interpreting the Stratigraphic Record*. New York: W. H. Freeman.

———. 2013. *Bringing Fossils to Life: An Introduction to Paleobiology*. 3rd ed. New York: Columbia University Press.

———. 2017. *Evolution: What the Fossils Say and Why It Matters*. 2nd ed. New York: Columbia University Press.

Prothero, Donald R., and Fred Schwab. 2013. *Sedimentary Geology*. 3rd ed. New York: W. H. Freeman.

Shipman, Pat. 1993. *Life History of a Fossil: An Introduction to Taphonomy and Paleoecology*. Cambridge, MA: Harvard University Press.

Chapter 12

Dalrymple, G. Brent. 1994. *The Age of the Earth*. Stanford, CA: Stanford University Press.

Hedman, Matthew. 2007. *The Age of Everything: How Science Explores the Past.* Chicago: University of Chicago Press.

Holmes, Arthur. 1913. *The Age of the Earth.* London: Harper and Brothers.

Isaak, Mark, 2006. *The Counter-Creationism Handbook.* Berkeley: University of California Press.

Kitcher, Philip. 1982. *Abusing Science: The Case against Creationism.* Cambridge, MA: MIT Press.

Lewis, Cherry. 2002. *The Dating Game: One Man's Search for the Age of the Earth.* Cambridge: Cambridge University Press.

Macdougall, Doug. 2008. *Nature's Clocks: How Scientists Measure the Age of Almost Everything.* Berkeley: University of California Press.

McGowan, Chris. 1984. *In the Beginning: A Scientist Shows Why the Creationists Are Wrong.* Buffalo, NY: Prometheus.

Montgomery, David R. 2012. *The Rocks Don't Lie: A Geologist Investigates Noah's Flood.* New York: W. W. Norton.

Numbers, Ronald. 1992. *The Creationists: The Evolution of Scientific Creationism.* New York: Knopf.

Prothero, Donald R. 1990. *Interpreting the Stratigraphic Record.* New York: W. H. Freeman.

———. 2017. *Evolution: What the Fossils Say and Why It Matters.* 2nd ed. New York: Columbia University Press.

———. 2018. *The Story of the Earth in 25 Rocks.* New York: Columbia University Press.

Prothero, Donald R., and Fred Schwab. 2013. *Sedimentary Geology.* 3rd ed. New York: W. H. Freeman.

Chapter 13

Christiansen, R. L., Calvert, A. T., and Grove T. L. 2017. *Geologic Field-Trip Guide to Mount Shasta Volcano, Northern California: U.S. Geological Survey Scientific Investigations Report 2017–5022-K3.* US Geological Survey. https://pubs.er.usgs.gov/publication/sir20175022K3.

Hill, Richard. 2004. *Volcanoes of the Cascades: Their Rise and Their Risks.* New York: Falcon Guides.

Moran, Mark, and Mark Sceurman. 2006. *Weird California.* San Francisco: Sterling.

Naef, Dustin W. 2018. *Mount Shasta's Forgotten History and Legends.* New York: Rive Fantasy.

Prothero, Donald R. 2018. *California's Amazing Geology.* New York: Taylor & Francis.

Zanger, Michael. 1995. *Mount Shasta: History, Legends, and Lore.* New York: Celestial Arts.

Chapter 14

Adams, Mark. 2015. *Meet Me in Atlantis: My Obsessive Quest to Find the Sunken City.* New York: Dutton.

Castleden, Rodney. 1993. *Minoans: Life in Bronze Age Crete.* London: Routledge.

———. 2001. *Atlantis Destroyed*. London: Routledge.

Charles River Editors. 2013. *The World's Greatest Civilizations: The History and Culture of the Minoans*. Self-published, CreateSpace.

———. 2016. *The Minoans and Myceneans: The History of Civilizations That First Developed Ancient Greek Culture*. Self-published, CreateSpace.

Feder, K. L. 1999. *Frauds, Myths, and Mysteries: Science and Pseudoscience in Archaeology*. Mountain View, CA: Mayfield.

Fouque, Ferdinand. 1999. *Santorini and Its Eruptions*. Baltimore, MD: Johns Hopkins University Press.

Friedrich, Walter. 2009. *Santorini: Volcano, Natural History, Mythology*. Aarhus: Aarhus University Press.

Friedrich, Walter, and Alexander McBirney. 2000. *Fire in the Sea: The Santorini Volcano: Natural History and the Legend of Atlantis*. Cambridge: Cambridge University Press.

Kershaw, Steve. 2018. *The Search for Atlantis: A History of Plato's Ideal State*. New York: Pegasus.

Pellegrino, Charles. 2017. *Unearthing Atlantis: An Archeological Odyssey to the Fabled Lost Civilization*. Self-published, CreateSpace.

Pellegrino, Charles, and Arthur C. Clarke. 1993. *Unearthing Atlantis: An Archeological Odyssey*. New York: Vintage.

Rethemiotakis, Giorgos. 2008. *From the Land of the Labyrinth: Minoan Crete, 3000–1000 BC*. Athens: Onassis Foundation.

Chapter 15

Devereux, Paul, and Ian Thomson. 1979. *The Ley Hunter's Companion: Aligned Ancient Sites: A New Study with Field Guide and Maps*. London: Thames and Hudson.

Watkins, Alfred. 1925. *The Old Straight Track: Its Mounds, Beacons, Moats, Sites, and Mark Stones*. London: Methuen & Co.

Chapter 16

Bonewitz, Ronald. 2008. *Rock and Gem: The Definitive Guide to Rocks, Minerals, Gemstones, and Fossils*. London: DK.

DK. 2012. *Smithsonian Nature Guide: Rocks and Minerals: The World in Your Hands*. London: DK.

National Audubon Society. 1979. *National Audubon Society Field Guide to Rocks and Minerals*. Washington, DC: National Audubon Society.

Symes, R. F. 2014. *DK Eyewitness Books: Crystal and Gem—Admire the Beauty and Versatility of Crystals and Gems*. London: DK.

Chapter 17

Agricola, Georgius. (1556) 1950. *De Re Metallica*. Translated by Herbert Hoover. New York and Basel: Dover.

Barrett, Linda K., and Evon Z. Vogt. 1969. "The Urban American Dowser." *Journal of American Folklore* 325: 195–213.

Barrett, William, and Theodore Besterman. (1926) 2004. *The Divining Rod: An Experimental and Psychological Investigation*. Reprint ed. Berlin: Kessinger.

Bird, Christopher. 1979. *The Divining Hand*. New York: Dutton.

Ellis, Arthur Jackson. 1917. *The Divining Rod: A History of Water Witching*. Washington, DC: US Government Printing Office.

Fiddick, Thomas. 2011. *Dowsing: With an Account of Some Original Experiments*. Sheffield, UK: Cornovia.

Gregory, John Walter. 1928. *Water Divining: Annual Report of the Smithsonian Institution*. Washington, DC: US Government Printing Office.

Randi, James. 1982. *Flim-Flam!* Buffalo, NY: Prometheus.

Chapter 18

Bartlett, Bruce. 2017. *The Truth Matters: A Citizen's Guide to Separating Facts from Lies, and Stopping Fake News in Its Tracks*. New York: Ten Speed.

Dunning, Brian. 2018. *Conspiracies Declassified: The Skeptoid Guide to the Truth behind the Theories*. New York: Adams Media.

Gardner, Martin. 1952. *Fads and Fallacies in the Name of Science*. New York: Dover.

———. 1981. *Science: Good, Bad, and Bogus*. Buffalo, NY: Prometheus.

Goldacre, Ben. 2010. *Bad Science: Quacks, Hacks, and Big Pharma Flacks*. New York: McClelland and Stewart.

Manjoo, Farhad. 2008. *True Enough: Learning to Live in a Post-Fact Society*. New York: Wiley.

Mooney, Chris. 2012. *The Republican Brain: The Science of Why They Deny Science—and Reality*. New York: Wiley.

Nichols, Tom. 2017. *The Death of Expertise: The Campaign against Established Knowledge and Why It Matters*. Oxford: Oxford University Press.

Pierce, Charles P. 2009. *Idiot America: How Stupidity Became a Virtue in the Land of the Free*. New York: Doubleday.

Prothero, Donald R. 2013. *Reality Check: How Science Deniers Threaten Our Future*. Bloomington: Indiana University Press.

Sagan, Carl. 1996. *The Demon-Haunted World: Science as a Candle in the Dark*. New York: Ballantine.

Shermer, Michael. 1997. *Why People Believe Weird Things: Pseudoscience, Superstition, and Other Confusions of Our Time*. New York: W. H. Freeman.

———. 2005. *Science Friction: Where the Known Meets the Unknown*. New York: Times Books.

———. 2011. *The Believing Brain: From Ghosts and Gods to Politics and Conspiracies—How We Construct Beliefs and Reinforce Them as Truths*. New York: Times Books.

Taibbi, Matt. 2008. *The Great Derangement: A Terrifying True Story of War, Politics, and Religion*. New York: Spiegel & Grau.

INDEX

DONALD R. PROTHERO taught college geology and paleontology for forty years, at Caltech, Columbia, and Cal Poly Pomona as well as Occidental, Knox, Vassar, Glendale, Mt. San Antonio, and Pierce Colleges. He is the author of numerous books and scientific papers, including *UFOS, Chemtrails, and Aliens: What Science Says*; *Reality Check: How Science Deniers Threaten Our Future*; and *Abominable Science!: Origins of the Yeti, Nessie, and Other Famous Cryptids*.